D0220678

INTRODUCTION TO
LINEAR ALGEBRA

INTRODUCTION TO

LINEAR ALGEBRA

Marvin Marcus

Professor of Computer Science
University of California, Santa Barbara

Henryk Minc

Professor of Mathematics
University of California, Santa Barbara

DOVER PUBLICATIONS, INC., NEW YORK

Copyright © 1965 by Marvin Marcus and Henryk Minc.
All rights reserved under Pan American and International Copyright Conventions.

Published in Canada by General Publishing Company, Ltd., 30 Lesmill Road, Don Mills, Toronto, Ontario.
Published in the United Kingdom by Constable and Company, Ltd.

This Dover edition, first published in 1988, is an unabridged, unaltered republication of the third, corrected printing (1969) of the work first published by The Macmillan Company, New York, 1965 (as "A mathematics text under the editorship of Carl B. Allendoerfer").

Manufactured in the United States of America
Dover Publications, Inc., 31 East 2nd Street, Mineola, N.Y. 11501

Library of Congress Cataloging-in-Publication Data

Marcus, Marvin, 1927–
    Introduction to linear algebra / by Marvin Marcus and Henryk Minc.
        p.   cm.
    Reprint. Originally published: New York : Macmillan, c1965.
    Includes index.
    ISBN 0-486-65695-0 (pbk.)
    1. Algebras, Linear.   I. Minc, Henryk.   II. Title.
QA184.M373   1988
512'.5—dc19   87-27796
                                                            CIP

TO JEFFREY AND KAREN
AND TO RALPH AND TADEUSZ

Preface

Linear algebra is a subject of central importance in both pure and applied mathematics. The problems of inverting large-order matrices, finding characteristic roots, and solving linear equations are common to many physical, biological, and social sciences. For these and other reasons, the elementary parts of the subject have lately begun to appear in the senior high school and first-year university curriculum.

In this book we present an introduction to the fundamental concepts in linear algebra and matrix theory. The book is intended for a one-semester course at the sophomore or junior level. Our approach is first to introduce vector spaces and then to regard matrices as concrete representations of linear transformations on vector spaces. We are not wedded either to the "matrix" or "linear transformation" approach to this subject. For each result we have tried to select the method that is most comprehensible and informative. In certain rarefied circles, the appearance of a matrix in linear algebra is regarded with a combination of suspicion and contempt. The present authors regard this attitude as somewhat unfortunate. It is at about the same level as the graduate student who is unable to write a proof of the continuity of $f(x) = x$ without using quantifier calculus.

There are several topics at the end of Chapter 3 that are not altogether standard for a book at this level—for example, Gerŝgorin disks, the Courant-Fischer theorem, and Cauchy's inequalities for characteristic roots of submatrices. These items are presented in such a way that a student who has developed some technique and understanding should be able to make it through them successfully. Moreover, these results begin to approach the periphery of research activity in some parts of the subject.

The prerequisite for a course given from this book is the successful completion of a standard course in what is usually called college algebra. In other words, the student should have a nodding acquaintance with complex numbers, polynomials, summation of finite series, and so forth.

We have not included the elementary divisor theory or any of the resulting canonical forms for matrices over rings of integers and polynomials. We have found it virtually impossible to include the Frobenius and Jordan normal forms in a one-semester course. It is also our opinion that the theory of elementary divisors is somewhat sophisticated for students who have not had the usual undergraduate course in modern algebra. It is true that we introduce the notions of ring and field immediately, but we essentially use the language without getting involved in any of the difficult and deep results concerning these concepts.

In general, we hope that this book will provide a quick penetration into some important and respectable mathematics certainly accessible to the average undergraduate student majoring in any of the sciences.

The authors wish to thank Professor B. N. Moyls for his many helpful suggestions. We also wish to express thanks to our long-suffering typist, Rebecca Michael.

Marvin Marcus
Henryk Minc

Contents

3

Characteristic Roots   143

Answers and Solutions   210
Index   257

CHAPTER 1

Vector Spaces and Linear Transformations

1.1 Vector Spaces

Linear algebra is the study of objects called vector spaces and certain operations on vector spaces called linear transformations. A vector space is a conglomerate of several things. The first of these is a set of numbers R, the second is a set of items V called vectors, the third is a way of combining pairs of elements in V and the fourth is a way of combining elements of R and V.

We begin our development with the definition of a ubiquitous mathematical object called a ring. Rings are generalizations of many familiar number systems that the reader has encountered before. Examples of rings abound: the real numbers; the rational numbers (i.e., fractions a/b where a and b are integers); the ordinary integers; the set of ordinary even integers. In every case the operations are addition and multiplication. As another example of a ring we may consider all polynomials $c_n x^n + c_{n-1} x^{n-1} + \cdots + c_1 x + c_0$ in which the coefficients c_i are integers and the operations are the usual ones. In the development of the subject matter of this book we will introduce many other more exotic examples of rings.

1

Definition 1.0 **(Ring)**   *A set R together with two operations called addition,* $+$, *and multiplication,* $\times$, *is called a ring if for every a, b, and c in R*

(i) $a + b$ *and* $a \times b$ *are uniquely defined elements of R;*

(ii) $(a + b) + c = a + (b + c)$, $a \times (b \times c) = (a \times b) \times c$ **(Associative laws);**

(iii) $a + b = b + a$ **(Commutativity of addition);**

(iv)   *there exists a unique element in R called zero and denoted by 0 such that* $a + 0 = 0 + a = a$ **(Existence of additive identity);**

(v)   *there is a unique element denoted by* $-a$ *such that* $a + (-a) = (-a) + a = 0$ **(Additive inverse);**

(vi) $a \times (b + c) = (a \times b) + (a \times c)$; $(b + c) \times a = (b \times a) + (c \times a)$ **(Distributive laws).**

A ring R is a field if in addition to the above axioms

(vii) $a \times b = b \times a$ **(Commutativity of multiplication);**

(viii) *there exists a unique element in R called the identity and denoted by 1 such that* $a \times 1 = 1 \times a = a$ **(Existence of multiplicative identity);**

(ix)   *if* $a \neq 0$ *then there exists a unique element denoted by* a^{-1} *such that* $a \times a^{-1} = a^{-1} \times a = 1$; a^{-1} *is called the* **inverse** *of a* **(Multiplicative inverse).**

If R is a ring in which (vii) and (viii) hold it should startle no one that R is called a **commutative ring** with a **multiplicative identity** 1.

For example, the set of complex numbers $a + ib$, a and b real, forms a field with the usual definitions of addition and multiplication. Both the real numbers and the rational numbers with the standard operations are also fields. A somewhat more unusual example is given by the following two tables for $+$ and $\times$.

$$
R: \quad
\begin{array}{c|ccc}
+ & a & b & c \\
\hline
a & a & b & c \\
b & b & c & a \\
c & c & a & b \\
\end{array}
\qquad
\begin{array}{c|ccc}
\times & a & b & c \\
\hline
a & a & a & a \\
b & a & b & c \\
c & a & c & b \\
\end{array}
\tag{1}
$$

The product $a \times b$ in a ring is generally denoted by ab, i.e., the times sign is left out.

Definition 1.1   **(Vector space)**   *A vector space V (over a field R) is a set of objects called* **vectors** *together with two operations. The first* *operation is called* **addition** *of vectors and for every u,v,w in V*

  (i) $u + v$ *is a uniquely defined element of* V,
 (ii) $u + (v + w) = (u + v) + w$,
(iii) $u + v = v + u$,
 (iv)   *there exists a vector 0_V such that $u + 0_V = 0_V + u = u$;*
  (v)   *for each u in V there exists a unique vector $-u$ such that $u + (-u) = (-u) + u = 0_V$.*

In other words, to each pair of elements u,v in V we associate another element of V called the *sum* of u and v, $u + v$. The rules governing addition of vectors look very much like the rules governing ordinary real number addition.

There also exists an operation between elements of R and elements of V called **scalar multiplication** *such that for all a,b in R and all u,v in V*

  (vi)   *au is a vector in V;*
 (vii) $a(u + v) = au + av$;
(viii) $(a + b)u = au + bu$;
  (ix) $a(bu) = (ab)u$;
   (x) $1u = u$, $0u = 0_V$.

The elements of R are called *scalars*. A vector space consisting of the single vector 0_V is called the zero vector space; all other vector spaces are said to be nonzero. We have denoted the scalar product au, $a \in R$, $u \in V$, by juxtaposing the two items. Also, the addition in V has the same notation, namely "+", as the addition in R. To add to the confusion we shall denote the zero vector, 0_V, by simply dropping the V to obtain 0. This confusion of symbols is more apparent than real and in any given context the meaning should be clear.

A familiar example of a vector space is the set of all "vectors" $u = (a_1, a_2)$, where a_1, a_2 are real numbers. The pictorial representation of u is an arrow in the plane whose tail is at the origin and whose head is at the point (a_1, a_2).

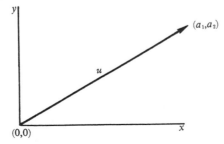

The addition is given by $u + v = (a_1, a_2) + (b_1, b_2) = (a_1 + b_1, a_2 + b_2)$ and the scalar multiplication by $au = (aa_1, aa_2)$. A reasonable generalization of this example is the vector space of n-tuples.

Definition 1.2 (*Vector space of n-tuples*) *For the set of all n-tuples $u = (a_1, \ldots, a_n)$, where the a_i are elements of a field R, define addition and scalar multiplication as follows:*

$$(a_1, \ldots, a_n) + (b_1, \ldots, b_n) = (a_1 + b_1, \ldots, a_n + b_n); \quad (2)$$
$$a(a_1, \ldots, a_n) = (aa_1, \ldots, aa_n). \quad (3)$$

This vector space is called the space of n-tuples over R and is denoted by $V_n(R)$. The zero in $V_n(R)$ is the n-tuple $(0, \ldots, 0)$. It is a simple task to verify that $V_n(R)$ does in fact satisfy the rules (i)–(x) in Def. 1.1.

As another example of a vector space let V be the set of all continuous real valued functions defined on the closed interval $0 \le x \le 1$. The addition for V is just ordinary function addition: $(f + g)(x) = f(x) + g(x)$. The scalar field R is taken to be the field of real numbers. If $a \in R$ and $f \in V$ then scalar multiplication is defined by $(af)(x) = a(f(x))$.

Definition 1.3 (*Linear dependence*) *If V is a vector space over R and $u_1, \ldots, u_k$ and $a_1, \ldots, a_k$ are respectively k vectors in V and k scalars in R then the vector u given by*

$$u = a_1 u_1 + a_2 u_2 + \cdots + a_k u_k \quad (4)$$

is called a **linear combination** *of $u_1, \ldots, u_k$ with* **coefficients** *$a_1, \ldots, a_k$. The totality of linear combinations of $u_1, \ldots, u_k$ obtained by allowing $a_1, \ldots, a_k$ to vary over R is a vector space. For, if c is any scalar, then*

$$c(a_1 u_1 + \cdots + a_k u_k) = (ca_1)u_1 + \cdots + (ca_k)u_k$$

is a linear combination of $u_1, \ldots, u_k$ and the sum of two linear combinations of $u_1, \ldots, u_k$,

$$(a_1 u_1 + \cdots + a_k u_k) + (b_1 u_1 + \cdots + b_k u_k)$$
$$= (a_1 + b_1)u_1 + \cdots + (a_k + b_k)u_k,$$

*is a linear combination of $u_1, \ldots, u_k$. The rest of the axioms in Def. 1.1 may be easily verified. The space of all linear combinations of vectors $u_1, \ldots, u_k$ is said to be **spanned** by $u_1, \ldots, u_k$ and this space is denoted by $\langle u_1, \ldots, u_k \rangle$. It may be that some of the u_j's are redundant in the sense that $\langle u_1, \ldots, u_k \rangle$ can be also spanned by fewer than all of $u_1, \ldots, u_k$. We shall show that this can happen if and only if a linear combination of $u_1, \ldots, u_k$ with at least one nonzero coefficient, is equal to 0. In order to do this we introduce an important new concept. If the only linear combination of $u_1, \ldots, u_k$ equal to 0 is the one for which all coefficients are zero then $u_1, \ldots, u_k$ are said to be **linearly independent**; otherwise they are **linearly dependent**. It is obvious that if $u_{i_1}, \ldots, u_{i_r}$ are some of the vectors of a linearly independent set of vectors $u_1, \ldots, u_k$ then $u_{i_1}, \ldots, u_{i_r}$ are also linearly independent. For if $c_{i_1}u_{i_1} + \cdots + c_{i_r}u_{i_r} = 0$ and not all of c_{i_j} are 0, then by setting $c_t = 0, t \neq i_1, \ldots, i_r$, we see that $c_1u_1 + \cdots + c_ku_k = 0$ and not all of $c_1, \ldots, c_k$ are 0. If there exists a finite set of vectors $u_1, \ldots, u_k$ such that $V = \langle u_1, \ldots, u_k \rangle$, then V is said to be **finite dimensional**. If $u_1, \ldots, u_k$ are linearly independent and span V then this set of vectors is called a **basis** of V.*

For example, $V_n(R)$ is finite dimensional. For, let

$$e_t = (\delta_{1t}, \delta_{2t}, \ldots, \delta_{nt}), \quad t = 1, \ldots, n,$$

where

$$\delta_{ij} - \begin{cases} 1 & \text{if } i = j, \\ 0 & \text{if } i \neq j, \end{cases}$$

is the *Kronecker delta*. If u is any n-tuple over R then

$$u = (a_1, \ldots, a_n) = a_1(1, 0, \ldots, 0) + a_2(0, 1, 0, \ldots, 0)$$
$$+ \cdots + a_n(0, \ldots, 0, 1)$$
$$= a_1e_1 + a_2e_2 + \cdots + a_ne_n.$$

Moreover, if $u = 0$ then $a_1 = \cdots = a_n = 0$ and hence $e_1, \ldots, e_n$ are linearly independent and thus form a basis for $V_n(R)$. We shall call this the *standard basis* for $V_n(R)$.

We remark that if $u_1, \ldots, u_k$ is a basis of V and

$$\sum_{s=1}^{k} c_s u_s = \sum_{s=1}^{k} d_s u_s$$

then $c_s = d_s$, $s = 1, \ldots, k$. For, $\displaystyle\sum_{s=1}^{k} (c_s - d_s)u_s = 0$ and hence $c_s - d_s = 0$, $s = 1, \ldots, k$.

In general, we shall use the capital sigma notation to denote linear combinations. Thus (4) becomes

$$u = \sum_{i=1}^{k} a_i u_i \qquad (5)$$

If X is any subset of V then $\langle X \rangle$ will denote the totality of linear combinations of vectors in X. The set $\langle X \rangle$ is called the *space* spanned by X. Each element of $\langle X \rangle$ is contained in V. We designate this by $\langle X \rangle \subset V$.

Theorem 1.1   *Let V be a finite dimensional vector space over the field R.*

  (i)   *If $u_1, \ldots, u_n$ are linearly dependent then some u_k is a linear combination of the rest of the u_i, $i \neq k$.*

 (ii)   *If V is nonzero (i.e., $V \neq \{0_V\}$) then V has a basis.*

(iii)   *If $u_1, \ldots, u_n$ is a basis of V and $v_1, \ldots, v_r$ are linearly independent in V then $r \leq n$ and for some set of $n - r$ of the u_i, say $u_{r+1}, \ldots, u_n$, the set $v_1, \ldots, v_r, u_{r+1}, \ldots, u_n$ is a basis of V.*

(iv)   *Any two bases of V contain the same number of vectors.*

Proof.   (i)   Suppose $c_1 u_1 + \cdots + c_n u_n = 0$ and some $c_i \neq 0$, say $c_k \neq 0$. Then

$$c_k u_k = -\sum_{i=1, i \neq k}^{n} c_i u_i, \qquad (6)$$

and since $c_k \neq 0$,

$$u_k = \sum_{i=1, i \neq k}^{n} (-c_k^{-1} c_i)u_i,$$

and we have expressed u_k as a linear combination of the rest of the u_i, $i \neq k$.

(ii)   Since V is nonzero and finite dimensional there exists a finite spanning set for V; call it X. Since X is finite it has only a finite number of subsets. In fact, it must contain a subset in which the vectors are linearly independent and which contains at least as many vectors as any other linearly independent subset. Let this maximal linearly independent subset of X be $u_1, \ldots, u_n$. Then clearly $\langle u_1, \ldots, u_n \rangle \subset \langle X \rangle$. If $x \in X$ then the set

$$x, u_1, \ldots, u_n \tag{7}$$

consists of $n + 1$ vectors and, from the maximality of n, the vectors must be linearly dependent. Hence there exist scalars $c_0, c_1, \ldots, c_n$, not all 0, such that

$$c_0 x + c_1 u_1 + \cdots + c_n u_n = 0. \tag{8}$$

If $c_0 = 0$ then (8) would imply that $u_1, \ldots, u_n$ were linearly dependent. Hence $c_0 \neq 0$ so we can write

$$x = \sum_{j=1}^{n} (-c_0^{-1} c_j u_j) \quad \in \langle u_1, \ldots, u_n \rangle. \tag{9}$$

We have proved that any x in the finite spanning set X is in $\langle u_1, \ldots, u_n \rangle$. Now let the elements of X be labeled $x_1, \ldots, x_p$. If v is any vector in V then $v \in \langle X \rangle = \langle x_1, \ldots, x_p \rangle$, i.e., $V = \langle X \rangle$. On the other hand, each x_t is in $\langle u_1, \ldots, u_n \rangle$ and, hence, we can write

$$x_t = \sum_{j=1}^{n} c_{tj} u_j, \quad t = 1, \ldots, p. \tag{10}$$

Finally, any $x \in \langle X \rangle$ can be written as

$$x = \sum_{t=1}^{p} a_t x_t,$$

where the coefficients a_t are in R.

By (10), we have

$$\sum_{t=1}^{p} a_t x_t = \sum_{t=1}^{p} a_t \sum_{j=1}^{n} c_{tj} u_j \tag{11}$$

$$= \sum_{j=1}^{n} \left(\sum_{t=1}^{p} a_t c_{tj} \right) u_j$$

$$\in \langle u_1, \ldots, u_n \rangle.$$

Hence $\langle X \rangle \subset \langle u_1, \ldots, u_n \rangle$. We now have

$$V = \langle X \rangle \subset \langle u_1, \ldots, u_n \rangle \subset \langle X \rangle.$$

In other words,

$$V = \langle u_1, \ldots, u_n \rangle, \tag{12}$$

the u_i are linearly independent, and thus they form a basis for V.

(iii)   We prove this first for $r = 1$. Consider the set of vectors

$$v_1, u_1, \ldots, u_n.$$

These are linearly dependent because $u_1, \ldots, u_n$ is a basis and v_1 is a linear combination of them. Thus there exist scalars $c_0, d_1, \ldots, d_n$, not all zero, such that

$$c_0 v_1 + d_1 u_1 + \cdots + d_n u_n = 0. \tag{13}$$

If every d_i in (13) were 0 it would follow that $v_1 = 0$, contradicting the hypothesis. Hence some d_i, which by a renumbering of the u's can be taken to be d_1, is not 0. Also $c_0 \neq 0$, otherwise $u_1, \ldots, u_n$ would be linearly dependent. We can then write

$$u_1 \in \langle v_1, u_2, \ldots, u_n \rangle.$$

It is also clear that $v_1, u_2, \ldots, u_n$ is a spanning set for V. Suppose that these vectors were linearly dependent so that

$$a_1 v_1 + a_2 u_2 + \cdots + a_n u_n = 0.$$

If a_1 were 0 it would follow that $u_2, \ldots, u_n$ would be linearly dependent, which is not the case. Hence,

$$v_1 = \sum_{j=2}^{n} (-a_1^{-1} a_j) u_j. \tag{14}$$

From (13) and (14) we have

$$\sum_{j=1}^{n} (c_0^{-1} d_j) u_j = \sum_{j=2}^{n} (a_1^{-1} a_j) u_j,$$

or

$$c_0^{-1}d_1u_1 + \sum_{j=2}^{n} (c_0^{-1}d_j - a_1^{-1}a_j)u_j = 0. \tag{15}$$

The linear independence of $u_1, \ldots, u_n$ then implies from (15) that $c_0^{-1}d_1 = 0$. But d_1 and c_0^{-1} are both nonzero. This whole house of cards has collapsed because we have assumed that $v_1, u_2, \ldots, u_n$ are linearly dependent. Thus they are linearly independent, span V, and therefore are a basis of V. This completes the proof for $r = 1$. The proof proceeds by induction. Assume the result true for $r - 1$ vectors $v_1, \ldots, v_{r-1}$. Then $r - 1 \leq n$. If $r - 1$ were equal to n then, by the induction hypothesis,

$$v_1, \ldots, v_{r-1}$$

would be a basis of V. This would mean that

$$v_r \in \langle v_1, \ldots, v_{r-1} \rangle$$

which contradicts the linear independence of $v_1, \ldots, v_r$. Thus $r - 1 < n$ and $r \leq n$. By the induction hypothesis we can assume that $v_1, \ldots, v_{r-1}$ can replace a set of $r - 1$ of the vectors $u_1, \ldots, u_n$, say $u_1, \ldots, u_{r-1}$, and the resulting set

$$v_1, \ldots, v_{r-1}, \quad u_r, u_{r+1}, \ldots, u_n \tag{16}$$

will be a basis.
Then

$$v_r = \sum_{j=1}^{r-1} a_j v_j + \sum_{j=r}^{n} b_j u_j, \tag{17}$$

where the a_j and b_j are in R. Since $v_1, \ldots, v_r$ are linearly independent, not all $b_j = 0$. Hence, by reordering the vectors $u_r, \ldots, u_n$ if necessary, we can assume $b_r \neq 0$. Solving (17) for u_r, we get

$$u_r = -\sum_{j=1}^{r-1} (b_r^{-1}a_j)v_j + b_r^{-1}v_r - \sum_{j=r+1}^{n} (b_r^{-1}b_j)u_j. \tag{18}$$

Thus

$$V = \langle v_1, \ldots, v_{r-1}, \; u_r, \; \ldots, \; u_n \rangle$$

$$\subset \langle v_1, \ldots, v_r, \; u_{r+1}, \; \ldots, \; u_n \rangle \subset V$$

and

$$v_1, \ldots, v_r, \; u_{r+1}, \; \ldots, \; u_n \tag{19}$$

span V. It remains to show that the vectors in (19) are linearly independent. Suppose that they are linearly dependent and that there exist scalars $d_1, \ldots, d_r$ and $c_{r+1}, \ldots, c_n$, not all 0, such that

$$\sum_{j=1}^{r-1} d_j v_j + d_r v_r + \sum_{j=r+1}^{n} c_j u_j = 0.$$

Now $d_r \neq 0$, otherwise the basis (16) would be linearly dependent. Hence

$$v_r = \sum_{j=1}^{r-1} (-d_r^{-1} d_j) v_j + \sum_{j=r+1}^{n} (-d_r^{-1} c_j) u_j. \tag{20}$$

Subtracting (20) from (17), we have

$$\sum_{j=1}^{r-1} (a_j + d_r^{-1} d_j) v_j + b_r u_r + \sum_{j=r+1}^{n} (b_j + d_r^{-1} c_j) u_j = 0. \tag{21}$$

Equation (21) is a linear dependence relation among the elements of the basis (16) and hence $b_r = 0$. However we know that $b_r \neq 0$ and thus the original assumption that the vectors in (19) are linearly dependent is untenable. It follows that (19) is a basis and (iii) is proved.

   (iv)   Suppose $u_1, \ldots, u_n$ and $v_1, \ldots, v_r$ are two bases of V. According to (iii), $r \leq n$ and by reversing the roles of the u_i's and v_i's it follows that $n \leq r$. Hence $r = n$.

Definition 1.4 (**Dimension**) *If V is a finite dimensional vector space over R then the common value of the number of vectors in any basis of V is called the **dimension** of V. This integer is denoted by $\dim V$. In case V consists of the zero vector only we set $\dim V = 0$.*

For example, we saw that $V_n(R)$ possesses the standard basis e_1, . . . , e_n. Hence

$$\dim V_n(R) = n. \tag{22}$$

In the proof of Theorem 1.1 we used a concept of "maximal" set several times. We next give this idea a formal definition.

Definition 1.5 (*Subspace, maximal sets*)   *If V is a vector space over R and W is a subset of V that is also a vector space over R, using the addition and scalar multiplication of V, then W is called a subspace of V. A set of p linearly independent vectors in a subset X of V will be called maximal in X if no linearly independent subset of X contains more than p vectors.*

As an example of a subspace let $V = V_3(R)$, where R is the field of real numbers. Let W be the subset of V consisting of all those $u = (u_1, u_2, u_3)$ for which $u_1 = u_2$. Then W is a subspace of V. For we need only show that W is closed under addition and scalar multiplication (all the other axioms for a vector space follow directly because W is a subset of V). But if $v = au = (au_1, au_2, au_3)$ and $u_1 = u_2$, then $au_1 = au_2$, so $v \in W$. Similarly, $u + w \in W$ when u and w are in W.

In the following results we exclude the trivial consideration of the space of dimension 0.

Theorem 1.2   *If V is a finite dimensional vector space over R and $\dim V = n$ then any $n + 1$ vectors in V are linearly dependent.*

Proof.   According to Theorem 1.1 (ii), (iv), V possesses a basis consisting of n vectors. Then Theorem 1.1 (iii) implies that if there is a set of r linearly independent vectors in V, then $r \le n$.

The structure of subspaces of a finite dimensional vector space is described in the following theorem.

Theorem 1.3   *Let V be a finite dimensional vector space over R and let W be a subspace of V. Then W is finite dimensional and*

$$\dim W \le \dim V. \tag{23}$$

Moreover, (23) is equality if and only if $W = V$.

Proof. The argument used to establish this result actually shows that a basis for W exists and any such basis can be augmented to obtain a basis of V. According to Theorem 1.2, no set of linearly independent vectors in V (and hence no set in W) can consist of more than dim $V = n$ vectors. Thus we can choose a maximal linearly independent set in W. Let $w_1, \ldots, w_r$ be such a maximal set in W and let $u_1, \ldots, u_n$ be a basis of V. According to Theorem 1.1 (iii), $r \leq n$. We assert that $w_1, \ldots, w_r$ is a basis of W. For, let $w \in W$ and consider the $r + 1$ vectors $w, w_1, \ldots, w_r$. These vectors are linearly dependent because of the maximality of r. Thus there exist $c_0, c_1, \ldots, c_r$, not all zero, such that

$$c_0 w + \sum_{i=1}^{r} c_i w_i = 0.$$

Now $c_0 \neq 0$, otherwise the set $w_1, \ldots, w_r$ would be linearly dependent. Hence

$$w = \sum_{i=1}^{r} (-c_0^{-1} c_i) w_i.$$

Thus $w_1, \ldots, w_r$ is a basis. Moreover, according to Theorem 1.1 (iii), there exist $n - r$ of the vectors $u_1, \ldots, u_n$, say $u_{r+1}, \ldots, u_n$, such that

$$w_1, \ldots, w_r, u_{r+1}, \ldots, u_n,$$

is a basis of V. In other words, any basis of W can be completed to a basis of V. Now (23) is equality if and only if $r = n$. In other words dim $W =$ dim V if and only if $w_1, \ldots, w_n$ is already a basis of V, i.e., $W = \langle w_1, \ldots, w_n \rangle = V$.

An important corollary follows from Theorem 1.3.

Theorem 1.4 *If V is a finite dimensional vector space over R and $X \subset V$ then a basis of the subspace $\langle X \rangle$ can be chosen from the elements of X.*

Proof. The set $\langle X \rangle$ is indeed a subspace of V, for linear combinations of vectors which are themselves linear combinations of vectors in X are once again linear combinations of vectors in X. The inclusions

$X \subset \langle X \rangle \subset V$ together with Theorem 1.3 show that no finite subset of X can contain more than $n = \dim V$ linearly independent vectors. Thus let $x_1, \ldots, x_p$ be a maximal linearly independent subset of X. If $x_0 \in X$ then the maximality of p implies that $x_0, x_1, \ldots, x_p$ are linearly dependent and thus $x_0 \in \langle x_1, \ldots, x_p \rangle$. In other words, $X \subset \langle x_1, \ldots, x_p \rangle \subset \langle X \rangle$. Further, any vector in $\langle X \rangle$ is a linear combination of vectors in X, ergo in $\langle x_1, \ldots, x_p \rangle$, and hence $\langle X \rangle \subset \langle x_1, \ldots, x_p \rangle$. Thus $\langle X \rangle = \langle x_1, \ldots, x_p \rangle$ and the vectors $x_1, \ldots, x_p$ constitute a basis of $\langle X \rangle$.

Quiz

Answer true or false:

1. If a and b are elements in a field R and $ab = 0$, then at least one of a or b is 0.
2. If the vector 0 is included among the vectors $v_1, \ldots, v_r$, then these vectors are linearly dependent.
3. A subset of a basis for a vector space consists of linearly independent vectors.
4. If u_1, u_2, u_3 are linearly dependent vectors in $V_4(R)$, then u_i is a scalar multiple of u_j, for some i and j, $i \neq j$.
5. The set of all vectors $x = (x_1, x_2) \in V_2(R)$ for which $x_1^2 - x_2^2 = 0$ is a subspace of $V_2(R)$. Take R to be the field of real numbers.
6. If X is a subset of a vector space V and $u_1, \ldots, u_r$ are in X, then $\langle u_1, \ldots, u_r \rangle \subset X$.
7. Any maximal set in a subspace W of a finite dimensional vector space V must be a basis of W.
8. A subspace of $V_3(R)$, where R is the real field, must always contain the origin.
9. The set of all solutions (x_1, x_2) to the linear equation $3x_1 + 2x_2 = 1$ comprises a subspace of $V_2(R)$, where R is the real field.
10. If X and Y are subsets of a vector space V and $\langle X \rangle = \langle Y \rangle = V$, then X and Y must overlap.

Exercises

1. Verify that the set R given in (1) is a field and identify the "0" and "1" elements.
2. Let V be the set of all infinite sequences of real numbers $(a_0, a_1, \ldots)$ in which at most a finite number of the a_i are different from 0. Define addition and scalar multiplication by

$$(a_0, a_1, \ldots) + (b_0, b_1, \ldots) = (a_0 + b_0, a_1 + b_1, \ldots)$$

and

$$r(a_0, a_1, \ldots) = (ra_0, ra_1, \ldots).$$

Prove that V is a vector space over the real numbers. Exhibit a basis. Is V finite dimensional?

3. In $V_3(R)$, where R is the real field, let X be the totality of vectors (x_1, x_2, x_3) for which $x_1^2 + x_2^2 - x_3^2 = 1$. What is the subspace $\langle X \rangle$? Find a basis in X for the subspace $\langle X \rangle$.

4. What are all possible subspaces of $V_2(R)$?

5. Show that the real field is a vector space over the rational field. Is it finite dimensional?

6. Show that if W is a subset of a vector space V over a field R such that $au + bv \in W$ for all $u, v \in W$ and $a, b \in R$ then W is a subspace of V.

1.2   Linear Transformations

The main business of algebraists is the study of structural relationships between algebraic systems. This is done by examining the properties of suitable functions defined on a given system with values in the same system or in another system of a similar type. In the case of vector spaces we study the properties of functions called linear transformations. Before we make the formal definition it is instructive to consider several examples.

Let V_n be the vector space of all polynomial functions $p(\lambda) = c_n\lambda^n + c_{n-1}\lambda^{n-1} + \cdots + c_1\lambda + c_0$ where $c_0, \ldots, c_n$ are real numbers. We assume that the degree of any polynomial in V_n is at most n. Although we have not discussed polynomial functions formally, these items have been with us in one form or another from grammar school days (see Sect. 3.1). Let D denote the operation of taking the derivative with respect to λ. Thus

$$Dp(\lambda) = p'(\lambda) = nc_n\lambda^{n-1} + (n-1)c_{n-1}\lambda^{n-2} + \cdots + c_1.$$

The elementary rules for differentiation state that

$$D[p(\lambda) + q(\lambda)] = D[p(\lambda)] + D[q(\lambda)]$$

and

$$D[cp(\lambda)] = cD[p(\lambda)]$$

for any two polynomials $p(\lambda)$ and $q(\lambda)$ and any scalar c. Observe that $Dp(\lambda) \in V_{n-1}$, i.e., the degree of the derivative of a polynomial in V_n is at most $n - 1$. The above two properties that make D interesting for our purposes can be combined in a single formula:

$$D[cp(\lambda) + dq(\lambda)] = c\, Dp(\lambda) + d\, Dq(\lambda)$$

for any scalars c,d and any polynomials $p(\lambda)$, $q(\lambda)$.

As another example consider the vector space $V_2(R)$ where R is the field of real numbers, and let T_θ denote the operation of rotating a vector $u = (a_1, a_2)$ counterclockwise through an angle θ. Thus if $v = T_\theta u$ then a picture of the action of T_θ looks as follows:

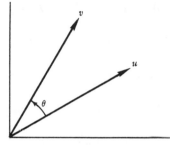

If we set $v = (b_1, b_2)$ then by elementary trigonometry

$$b_1 = a_1 \cos \theta - a_2 \sin \theta$$
$$b_2 = a_1 \sin \theta + a_2 \cos \theta.$$

These formulas for b_1 and b_2 in terms of a_1 and a_2 show immediately that

$$T_\theta(u + w) = T_\theta u + T_\theta w$$

and

$$T_\theta(cu) = cT_\theta u$$

for any vectors u and w and scalar c.

As a third example, let V be the vector space of all real-valued functions which are defined on the infinite interval $-\infty < x < \infty$. Let k be a fixed real number and define an operation T_k on V by

$$T_k[f(x)] = g(x)$$

where

$$g(x) = f(x + k) \qquad \text{for all } x.$$

Then clearly

$$T_k[f(x) + h(x)] = T_k[f(x)] + T_k[h(x)]$$

and

$$T_k[cf(x)] = cT_k[f(x)].$$

All three of the above examples involve a function defined on a vector space V which takes on values in a vector space. Moreover, if T designates any of the preceding functions then T satisfies

$$T(cu + dw) = cT(u) + dT(w)$$

for any two vectors u and w and any two scalars c and d. When convenient we shall delete the parentheses in $T(u)$ and simply write Tu. We generalize this property in the following important definition.

Definition 2.1 (**Linear transformation**) *Let U and V be vector spaces over the field R. Let T be a function defined on U with values in V which satisfies*

$$T(u_1 + u_2) = T(u_1) + T(u_2) \tag{1}$$

and

$$T(au) = a[T(u)] \tag{2}$$

*for all u_1, u_2, u in U and all $a \in R$. Then T is called a **linear transformation on U to V**.*

Clearly conditions (1) and (2) can be expressed as a single condition:

$$T(a_1 u_1 + a_2 u_2) = a_1 T(u_1) + a_2 T(u_2)$$

for all u_1, u_2 in U and all a_1, a_2 in R.

There are several terms used to designate T. Two of these are *linear mapping* and *homomorphism*.

In general, two functions defined on the same set are equal if their function values are equal. Thus two linear transformations T_1 and T_2 defined on U with values in V are equal if $T_1(u) = T_2(u)$ for all $u \in U$.

Definition 2.2 (*Combining transformations*) *Let T_1 and T_2 be linear transformations on U to V. Then the* **sum** *of T_1 and T_2 is the transformation S denoted by*

$$S = T_1 + T_2 \tag{3}$$

and defined by

$$S(u) = T_1(u) + T_2(u)$$

for all $u \in U$.

Similarly, we define the **scalar multiplication** *of a transformation T on U to V by an element $a \in R$. We set $M = aT$ where*

$$M(u) = a[T(u)]$$

for all $u \in U$.

If W is a third vector space over R and T_3 is a linear transformation on V to W then the **product** *of T_3 and T_1 is the transformation P denoted by*

$$P = T_3 T_1 \tag{4}$$

and defined by

$$P(u) = T_3[T_1(u)]$$

for each $u \in U$.

We first note that both the product and sum as defined above are linear transformations. For,

$$\begin{aligned}
S(u_1 + u_2) &= T_1(u_1 + u_2) + T_2(u_1 + u_2) \\
&= T_1(u_1) + T_1(u_2) + T_2(u_1) + T_2(u_2) \\
&= (T_1 + T_2)(u_1) + (T_1 + T_2)(u_2) \\
&= S(u_1) + S(u_2).
\end{aligned}$$

Similarly,

$$S(au) = aS(u)$$

for all $a \in R$ and all $u \in U$. To see that P is linear observe that

$$
\begin{aligned}
P(u_1 + u_2) &= T_3[T_1(u_1 + u_2)] \\
&= T_3[T_1(u_1) + T_1(u_2)] \\
&= T_3[T_1(u_1)] + T_3[T_1(u_2)] \\
&= P(u_1) + P(u_2).
\end{aligned}
$$

Similarly,

$$
P(au) = aP(u).
$$

We shall use the following notation: $L(U,V)$ will denote the totality of linear transformations on U to V. If $U = V$ we define the *identity* transformation I_U by $I_U(u) = u$ for all $u \in U$. It takes very little effort to verify that $I_U \in L(U,U)$. Moreover, if $T \in L(U,V)$ then $T = I_V T I_U$. If $T \in L(U,V)$ then we shall also indicate this with the notation

$$
T: U \to V. \tag{5}
$$

Let

$$
T_1: U \to V, \quad T_2: U \to V, \quad T_3: V \to W.
$$

By calculations almost identical to the ones given above we see that

$$
T_3(T_1 + T_2) = T_3 T_1 + T_3 T_2
$$

and if $T_4: X \to U$, then

$$
(T_1 + T_2)T_4 = T_1 T_4 + T_2 T_4.
$$

These are quite expectedly known as the *distributive laws* for combining linear transformations.

Definition 2.3   (*Kernel, range*)   *If $T: U \to V$, then there are two sets associated with T: the **kernel** of T is the totality of $u \in U$ for which $T(u) = 0 \in V$; the **range** of T is the totality of vectors of the form $T(u) \in V$. These are denoted respectively by* ker T *and* rng T.

As an example, let $T: V_3(R) \to V_2(R)$, where R is the field of real numbers and $T(a_1, a_2, a_3) = (a_2, a_3)$. Then the range of T is all of

$V_2(R)$ since any pair (a_2,a_3) is a value of T. Thus

$$\text{rng } T = V_2(R).$$

Also, if $T(a_1,a_2,a_3) = 0$ then $(a_2,a_3) = (0,0)$, i.e., $a_2 = a_3 = 0$. Thus $(a_1,a_2,a_3) \in \text{ker } T$ if and only if $a_2 = a_3 = 0$. We can then state that

$$\text{ker } T = \{u|u = ae_1,\ a \in R\} = \langle e_1 \rangle.$$

This "set maker" notation is read "the totality of u such that $u = ae_1$ for some $a \in R$." That is, ker T is the totality of multiples of the vector $e_1 = (1,0,0)$. Notice that in this example both rng T and ker T are vector spaces and moreover

$$\dim (\text{rng } T) + \dim (\text{ker } T) = \dim (V_3(R)),$$

i.e., $2 + 1 = 3$. These facts are generally true and we collect a number of important results together in the following theorem.

Theorem 2.1   *If U and V are finite dimensional vector spaces over a field R and $T \in L(U,V)$ then*
   (i)   *ker T is a subspace of U, rng T is a subspace of V and*

$$\dim (\text{ker } T) + \dim (\text{rng } T) = \dim U \tag{6}$$

   (Sylvester's law of nullity);
   (ii)   *$L(U,V)$, with addition and scalar multiplication as in Def. 2.2, forms a vector space over R and*

$$\dim L(U,V) = \dim U \dim V; \tag{7}$$

   (iii)   *if W and Y are vector spaces over R and $T_1\colon V \to W$, $T_2\colon W \to Y$ then*

$$(T_2T_1)T = T_2(T_1T). \tag{8}$$

Proof.   (i)   By Exercise 6, Sec. 1.1, it is sufficient to show that, for any u_1 and u_2 in ker T and $a_1,a_2 \in R$, the vector $a_1u_1 + a_2u_2 \in \text{ker } T$. We have

$$T(a_1u_1 + a_2u_2) = a_1T(u_1) + a_2T(u_2) = 0.$$

Hence ker T is a subspace. Similarly, if v_1 and v_2 are in rng T then there exist u_1, $u_2 \in U$ such that $T(u_1) = v_1$ and $T(u_2) = v_2$. We have

$$a_1 v_1 + a_2 v_2 = a_1 T(u_1) + a_2 T(u_2) = T(a_1 u_1 + a_2 u_2).$$

In other words, $a_1 v_1 + a_2 v_2$ is in the range of T. To prove (6) let $u_1, \ldots, u_r$ be a basis of ker T and complete to a basis of U with $u_{r+1}, \ldots, u_n$ (see Theorem 1.3, in particular the first part of the proof). If $v \in$ rng T then $v = T(u)$ for some $u \in U$. Let

$$u = \sum_{t=1}^{r} c_t u_t + \sum_{t=r+1}^{n} c_t u_t.$$

Then

$$v = T(u) = \sum_{t=1}^{r} c_t T(u_t) + \sum_{t=r+1}^{n} c_t T(u_t). \tag{9}$$

The first sum on the right in (9) must be 0 because $u_1, \ldots, u_r$ are all in ker T. Since v is any vector in rng T, we have

$$\text{rng } T = \langle T(u_{r+1}), \ldots, T(u_n) \rangle.$$

Suppose

$$\sum_{t=r+1}^{n} c_t T(u_t) = 0.$$

Then

$$T \left(\sum_{t=r+1}^{n} c_t u_t \right) = 0,$$

which in turn implies that

$$\sum_{t=r+1}^{n} c_t u_t \in \text{ker } T = \langle u_1, \ldots, u_r \rangle.$$

Thus for scalars $a_1, \ldots, a_r$

$$\sum_{t=r+1}^{n} c_t u_t = \sum_{t=1}^{r} a_t u_t$$

and the linear independence of $u_1, \ldots, u_n$ implies that $c_{r+1} = \cdots$ $= c_n = 0$. Hence $T(u_{r+1}), \ldots, T(u_n)$ do in fact constitute a basis of rng T. In other words,

$$n - r = \dim (\text{rng } T)$$

or

$$r + \dim (\text{rng } T) = n.$$

But, $r = \dim (\ker T)$ and $n = \dim U$, thus establishing (6).

(ii)   The addition of two linear transformations and the multiplication of a transformation by a scalar were defined in Def. 2.2. Now, define $-T$ by the formula $(-T)(u) = -T(u)$ and $0\colon U \to V$ by $0(u) = 0$. [It is amusing to observe that $0(0) = 0$ is a sensible formula where the first zero is the zero linear transformation, the second zero is 0_U and the third zero is 0_V]. The axioms in Def. 1.1 can be easily verified by the student (see Exercise 1).

To prove (7) we let $u_1, \ldots, u_n$ and $v_1, \ldots, v_m$ be bases of U and V respectively. For each pair i,j, $1 \leq i \leq n$, $1 \leq j \leq m$, let $T_{ij}\colon U \to V$ be defined as follows:

$$T_{ij}(u_s) = \delta_{is}v_j, \quad s = 1, \ldots, n, \tag{10}$$

$$T_{ij}\left(\sum_{s=1}^{n} a_s u_s \right) = \sum_{s=1}^{n} a_s T_{ij}(u_s). \tag{11}$$

Thus

$$T_{ij}\left(\sum_{s=1}^{n} a_s u_s \right) = \sum_{s=1}^{n} a_s \delta_{is} v_j$$

$$= a_i v_j.$$

This process of defining T_{ij} deserves a little conversation. Equation (10) gives the value of T_{ij} on each of the basis vectors $u_1, \ldots, u_n$. Formula (11) gives us the value of T_{ij} on any vector

$$u = \sum_{s=1}^{n} a_s u_s.$$

This procedure is called *linear extension* and always produces a linear transformation (see Exercise 2). We must show two things about the T_{ij}: first, they span $L(U,V)$ and second, they are linearly independent Thus, let $T \in L(U,V)$ and suppose

$$T(u_s) = \sum_{j=1}^{m} c_{sj}v_j, \quad s = 1, \ldots, n.$$

Then let S be the linear combination of the T_{ij} defined by

$$S = \sum_{i=1,j=1}^{n,m} c_{ij}T_{ij}.$$

We compute that

$$S(u_s) = \sum_{i=1,j=1}^{n,m} c_{ij}T_{ij}(u_s) \qquad (12)$$

$$= \sum_{i=1,j=1}^{n,m} c_{ij}\delta_{is}v_j$$

$$= \sum_{j=1}^{m} c_{sj}v_j$$

$$= T(u_s).$$

We have proved that $S(u_s) = T(u_s)$, $s = 1, \ldots, n$, and hence

$$S\left(\sum_{s=1}^{n} a_su_s \right) = \sum_{s=1}^{n} a_sS(u_s)$$

$$= \sum_{s=1}^{n} a_sT(u_s)$$

$$= T\left(\sum_{s=1}^{n} a_su_s \right).$$

To prove the linear independence of the T_{ij}, suppose that

$$\sum_{i=1,j=1}^{n,m} b_{ij}T_{ij} = 0,$$

then from (10) we have

$$0 = \sum_{i=1, j=1}^{m,n} b_{ij} T_{ij}(u_s) = \sum_{j=1}^{m} b_{sj} v_j, \quad s = 1, \ldots, n.$$

But $v_1, \ldots, v_m$ is a basis of V and hence these vectors are linearly independent. It follows that

$$b_{sj} = 0, \quad s = 1, \ldots, n, \quad j = 1, \ldots, m.$$

This means that the nm linear transformations T_{ij} are linearly independent, and thus

$$\dim L(U, V) = \dim U \dim V.$$

(iii)   The proof of (8) is trivial and depends only on the concept of function composition. For, if $u \in U$ then by definition

$$\begin{aligned}(T_2 T_1) T(u) &= T_2[T_1(T(u))] \\ &= T_2[T_1 T(u)] \\ &= T_2(T_1 T)(u).\end{aligned}$$

The associativity of multiplication of linear transformations established in Theorem 2.1 (iii) allows us to define unambiguously the product of more than two transformations. Thus

$$T_2 T_1 T = (T_2 T_1) T = T_2(T_1 T)$$

and

$$T_3 T_2 T_1 T = T_3(T_2 T_1 T), \text{ etc.}$$

If $V = U$, $T \in L(U, U)$ and k is a nonnegative integer, then T^k will denote the product $TT \cdots T$ in which T occurs as a factor k times. For convenience we will set $T^0 = I_U$. It is a very simple matter to verify that the usual laws of exponents hold:

$$\begin{aligned}T^p T^q &= T^{p+q}, \\ (T^p)^q &= T^{pq}.\end{aligned}$$

As an example of the use of Theorem 2.1 we can derive the fundamental existence theorem for solutions of systems of linear equations. Before we do this we shall make the following important definition.

Definition 2.4 *(Rank, nullity)* *Let U and V be finite dimensional vector spaces over a field R. If $T \in L(U,V)$ then the **rank** of T is the dimension of the range of T. This integer is denoted by $\rho(T)$. Thus*

$$\rho(T) = \dim (\text{rng } T).$$ (13)

The **nullity** *of T is the dimension of the kernel of T and is denoted by $\eta(T)$:*

$$\eta(T) = \dim (\text{ker } T).$$

Thus Sylvester's law of nullity given in (6) is translated into

$$\eta(T) + \rho(T) = \dim U;$$ (14)

and in words "nullity plus rank equals dimension of domain."

Now let a_{ij} and b_i, $i = 1, \ldots, m$, $j = 1, \ldots, n$, be numbers in the field R. The problem of *solving linear equations* is to determine the totality of vectors $x = (x_1, \ldots, x_n) \in V_n(R)$ such that

$$\sum_{j=1}^{n} a_{ij}x_j = b_i, \quad i = 1, \ldots, m.$$ (15)

Thus a typical problem could be: find all 3-tuples (x_1,x_2,x_3) such that

$$2x_1 + 2x_2 + 7x_3 = 1$$
$$x_1 - x_2 + 9x_3 = 0.$$

It is easy to show that a linear transformation $A: V_n(R) \rightarrow V_m(R)$ can be defined in terms of the numbers a_{ij}: for each $v \in V_n(R)$ define $y = Av$ by

$$y_i = \sum_{j=1}^{n} a_{ij}v_j, \quad i = 1, \ldots, m.$$

Theorem 2.2 *The system of linear equations (15) has a solution if and only if*

$$b = (b_1, \ldots, b_m) \in \text{rng } A.$$ (16)

*If $z \in V_n(R)$ is a particular solution to (15) then the set of all solutions to
(15) consists of precisely those vectors $w \in V_n(R)$ of the form*

$$w = z + u \tag{17}$$

where $u \in \ker A$.

Proof.   To say that there exists $x \in V_n(R)$ such that $Ax = b$ is precisely the statement that b is in the range of A. Thus (16) is equivalent
to the existence of a solution to (15). Now suppose w and z are solutions
to (15). Then $A(w - z) = Aw - Az = b - b = 0$. Hence, any two
solutions of (15) differ by a vector in $\ker A$. On the other hand, if
$u \in \ker A$, then

$$A(z + u) = Az + Au = Az.$$

Theorem 2.2 tells us that a procedure for finding all solutions to (15)
consists of two parts:
   (a)   find all solutions of the *homogeneous* system $Ax = 0$;
   (b)   find one solution of the *nonhomogeneous* system $Ax = b$.

In Chap. 2 we shall take up a systematic way of actually computing
the set of solutions to a general linear system.
   We can restate Theorem 2.2 in somewhat different terms.

Theorem 2.3   *The system of linear equations (15) has a solution if
and only if*

$$\rho(A) = \dim \langle \operatorname{rng} A, b \rangle. \tag{18}$$

Proof.   Recall that $\rho(A) = \dim (\operatorname{rng} A)$ and suppose that $\dim (\operatorname{rng} A) =
\rho(A) = \dim \langle \operatorname{rng} A, b \rangle$. Then, by Theorem 1.3,

$$\operatorname{rng} A = \langle \operatorname{rng} A, b \rangle. \tag{19}$$

Hence, $b \in \operatorname{rng} A$ and (15) has a solution by Theorem 2.2.
   Conversely if (15) has a solution then Theorem 2.2 states that
$b \in \operatorname{rng} A$. Hence (19) holds, which in turn implies (18).

   As an illustration of Theorem 2.2 consider the system of linear equations over the rational number field R:

$$2x_1 + 6x_2 - x_3 + x_4 = 2$$
$$x_1 + 3x_2 + x_3 \qquad = 5 \qquad\qquad (20)$$
$$-x_1 - 3x_2 - x_3 \qquad = 0.$$

We want to know if the vector $b = (2,5,0)$ is in the range of A. Consider the vectors $v_i = Ae_i$, $i = 1, \ldots, 4$, where the e_i are the standard basis vectors in $V_4(R)$. Any vector $x \in V_4(R)$ is of the form

$$x = \sum_{t=1}^{4} x_t e_t$$

and hence

$$Ax = \sum_{t=1}^{4} x_t A e_t = \sum_{t=1}^{4} x_t v_t.$$

In other words $v_1, \ldots, v_4$ span rng A. Now we compute immediately from (20) that

$$v_1 = Ae_1 = (2,1,-1)$$
$$v_2 = Ae_2 = (6,3,-3)$$
$$v_3 = Ae_3 = (-1,1,-1)$$
$$v_4 = Ae_4 = (1,0,0).$$

Observe that $v_2 = 3v_1$ and $v_3 + 3v_4 = v_1$. Hence

$$\text{rng } A = \langle v_1, v_4 \rangle.$$

The question of the existence of a solution to (20) is therefore equivalent to the question of the existence of rational numbers c_1 and c_2 for which

$$(2,5,0) = c_1(2,1,-1) + c_2(1,0,0)$$
$$= (2c_1 + c_2,\, c_1,\, -c_1).$$

But clearly the two conditions $c_1 = 5$, $-c_1 = 0$ are mutually incompatible. Hence (20) has no solutions.

Quiz

Answer true or false (in what follows U and V are finite dimensional vector spaces over a field R):

1. If $T \in L(U,U)$ then $T^2 = TT \in L(U,U)$.
2. If $T \in L(U,V)$ and $\eta(T) = \dim U$ then $T = 0$.
3. There are n^2 linearly independent linear transformations in $L(U,U)$ where $\dim U = n$.
4. If T and S are in $L(U,U)$, then $\ker S \subset \ker TS$.
5. If $U = V_3(R)$ there is no $T \in L(U,U)$ such that $T(e_1) = e_1$, $T(e_2) = e_3$, $T(e_3) = -e_3$ and $T(e_1 + e_2 + e_3) = 2e_1$.
6. If T and S are in $L(U,V)$ and $T(g_j) = S(g_j)$, $j = 1, \ldots, n$, where $\{g_1, \ldots, g_n\}$ is a basis of U, then $T = S$.
7. If S and T are in $L(U,U)$, then $(S + T)^2 = S^2 + 2ST + T^2$.
8. If $T \in L(U,U)$ and $\dim U = n$, then $\rho(T)\eta(T) \leq n^2/4$.
9. If $T \in L(U,U)$ and $T^2 = 0$, then $T = 0$.
10. If $T \in L(U,V)$ and $\ker T \neq 0$, then there exist vectors u_1 and u_2 in U such that $u_1 \neq u_2$ and $Tu_1 = Tu_2$.

Exercises

1. Complete the proof of Theorem 2.1 (ii) by verifying that $L(U,V)$ satisfies the axioms in Def. 1.1.
2. Show that the process of linear extension [see proof of Theorem 2.1 (ii)] always produces a linear transformation.
3. Show that the system of linear equations

$$2x_1 + x_2 - 3x_3 = 0$$
$$x_2 + x_3 - 7x_4 = -1$$
$$x_3 + x_4 = 0$$

has a solution.
4. Show that if $T \in L(U,V)$ and $\dim U > \dim V$ then there exists a non-zero vector $u \in U$ such that $Tu = 0$.
5. Write a correct formula involving four different "zeros" without using a "plus" sign.
6. Let S and T be in $L(U,V)$. Suppose that $u_1, \ldots, u_n$ is a basis of U. Show that $S = T$ if and only if $Su_s = Tu_s$, $s = 1, \ldots, n$.
7. Let $V = V_3(R)$ and let $T \in L(V,V)$ be defined by $Te_1 = e_2$, $Te_2 = e_3$, $Te_3 = 0$ where e_1, e_2, e_3 is the standard basis for $V_3(R)$. Show that $T \neq 0$, $T^2 \neq 0$ but $T^3 = 0$.
8. Compute the rank and nullity of the transformation T in the preceding exercise.

9. Consider the system of equations over the rational numbers R:

$$2x_1 + 6x_2 - x_3 + x_4 = 0$$
$$x_1 + 3x_2 + x_3 \qquad = 0$$
$$-x_1 - 3x_2 - x_3 \qquad = 0.$$

Find a nontrivial (i.e., not all of x_1, x_2, x_3, x_4 equal 0) solution in integers of the system.

1.3 Inner Product

In Sec. 1.1 we indicated how vectors in $V_2(R)$, where R is the real field, can be represented as directed line segments. This geometric interpretation allows us to speak about the length of a vector in $V_2(R)$ and about the angle between two nonzero vectors. There is a similar geometric interpretation for $V_3(R)$. The generalization of these geometric situations leads to the concept of an inner product which adds considerable variety and structure to the theory. Henceforth unless we specifically state that R is an arbitrary field, we shall assume that R is either the field of real numbers, Re, or the field of complex numbers, C.

In order to motivate the definition of inner product we consider the cosine of the angle between two vectors of unit length in $V_2(\text{Re})$:

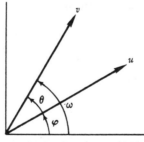

As we know from elementary trigonometry, if $u = (u_1, u_2)$ and $v = (v_1, v_2)$ then

$$\cos \theta = u_1 v_1 + u_2 v_2.$$

To prove this last formula, recall that if φ and ω denote the angles that u and v make with the positive horizontal axis respectively then

$$\cos \varphi = \frac{u_1}{\sqrt{u_1{}^2 + u_2{}^2}}, \quad \sin \varphi = \frac{u_2}{\sqrt{u_1{}^2 + u_2{}^2}}$$

$$\cos \omega = \frac{v_1}{\sqrt{v_1^2 + v_2^2}}, \quad \sin \omega = \frac{v_2}{\sqrt{v_1^2 + v_2^2}}.$$

But u and v are unit vectors so that $u_1^2 + u_2^2 = v_1^2 + v_2^2 = 1$. Now $\theta = \omega - \varphi$ and using the formula for $\cos (\omega - \varphi)$ we have

$$\cos \theta = \cos \omega \cos \varphi + \sin \omega \sin \varphi = u_1 v_1 + u_2 v_2.$$

In general, if u and v are arbitrary nonzero vectors, not necessarily of unit length, then

$$\cos \theta = \frac{u_1 v_1 + u_2 v_2}{\sqrt{u_1^2 + u_2^2} \sqrt{v_1^2 + v_2^2}}. \tag{1}$$

Let us designate the numerator of the fraction (1) by (u,v). We can think of (u,v) as simply a number associated with each pair of vectors u and v. Notice that $u_1^2 + u_2^2 = (u,u)$ so that we can rewrite (1) compactly as

$$\cos \theta = \frac{(u,v)}{(u,u)^{\frac{1}{2}}(v,v)^{\frac{1}{2}}}. \tag{2}$$

Also observe that if either of u or v is 0, then $(u,v) = 0$. Moreover (u,v) is linear in each of u and v separately. If u and v are perpendicular then θ is a right angle, $\cos \theta = 0$, and hence $(u,v) = 0$. We are thus led to the definition of an inner product in a vector space.

Definition 3.1 (**Inner product**) *Let V be a vector space over R where R is either the complex or the real field. Then an **inner product** on V is a function (u,v) on ordered pairs of vectors u,v in V with values in R satisfying:*

(i) (**conjugate symmetric**) *$(u,v) = \overline{(v,u)}$ for all u,v (the bar indicates complex conjugate);*

(ii) (**linearity**) *$(au_1 + bu_2, v) = a(u_1,v) + b(u_2,v)$ for all u_1,u_2,v in V and all a and b in R;*

(iii) (**positive-definite**) *for any vector u, $(u,u) \geq 0$ and $(u,u) = 0$ if and only if $u = 0$.*

We can immediately compute from (i) and (ii) that

$$(u, av_1 + bv_2) = \overline{(av_1 + bv_2, u)}$$
$$= \overline{a(v_1,u) + b(v_2,u)}$$
$$= \bar{a}\overline{(v_1,u)} + \bar{b}\overline{(v_2,u)}$$
$$= \bar{a}(u,v_1) + \bar{b}(u,v_2).$$

Thus

$$(u, av_1 + bv_2) = \bar{a}(u,v_1) + \bar{b}(u,v_2). \tag{3}$$

The two properties (ii) and (3) together are called *conjugate bilinearity*.

The most familiar inner product is the straightforward generalization of the example immediately preceding Def. 3.1.

Definition 3.2   (*Standard inner product*)   *Let R be either the real or the complex field, let V be a finite dimensional vector space over R, and suppose that $v_1, \ldots, v_n$ is a basis of V. For each pair of vectors*

$$u = \sum_{i=1}^{n} c_i v_i, \quad v = \sum_{i=1}^{n} d_i v_i$$

define

$$(u,v) = \sum_{i=1}^{n} c_i \bar{d}_i. \tag{4}$$

Then (u,v) is the **standard inner product** *relative to the basis $v_1, \ldots, v_n$. If $V = V_n(R)$, then the standard inner product relative to the standard basis $e_1, \ldots, e_n$ will simply be called the standard inner product for $V_n(R)$.*

Theorem 3.1   *The standard inner product relative to a basis $v_1, \ldots, v_n$ in V is an inner product.*

Proof.   Any vector is uniquely representable as a linear combination of the basis vectors $v_1, \ldots, v_n$ and hence formula (4) defines a function on pairs of vectors u,v in V. We verify (iii) as typical of the kind of calculations needed here and leave the verification of (i) and (ii) as an exercise. Thus

$$(u,u) = \sum_{i=1}^{n} c_i \bar{c}_i = \sum_{i=1}^{n} |c_i|^2 \geq 0.$$

This latter inequality can be equality if and only if every c_i is zero, i.e., $u = 0$.

Our next definition introduces some standard nomenclature.

Definition 3.3   (*Nomenclature for inner product*)   *Let V be a vector space over R where R is either the real or the complex field. If*

(u,v) *is an inner product on* V, *then* V *together with the inner product is called a* **unitary space.** *If* R *happens to be the real field then the term is* **Euclidean space.** *If a pair of vectors* u,v *satisfies* $(u,v) = 0$ *then* u *and* v *are* **perpendicular** *or* **orthogonal.** *If* $u_1, \ldots, u_p$ *is a set of vectors for which* $(u_i,u_j) = 0$ *when* $i \neq j$ *then the vectors are said to be* **mutually orthogonal.** *If* u *is any vector then the non-negative real number* $(u,u)^{\frac{1}{2}}$ *is called the* **norm** *or* **length** *of* u. *The norm is denoted by* $\|u\| = (u,u)^{\frac{1}{2}}$. *If* $u_1, \ldots, u_p$ *are mutually orthogonal vectors of unit length then they are called* **orthonormal,** *abbreviated* o.n. *A single vector of unit length is called a* **unit** *vector. If* L *is any set of vectors in* V *then the totality of vectors* $x \in V$ *for which* $(x,y) = 0$ *for every* $y \in L$ *is called the* **orthogonal complement** *of* L. *We designate this set by* $L^{\perp}$.

If (u,v) is the standard inner product relative to a basis $v_1, \ldots, v_n$ then the definition given in (4) yields

$$(v_i,v_j) = \delta_{ij}, \quad i,j = 1, \ldots, n. \tag{5}$$

In other words, $v_1, \ldots, v_n$ is an o.n. basis with respect to the inner product. Thus the standard basis of $V_n(R)$ is an o.n. basis with respect to the standard inner product. It is easy to construct other o.n. bases in $V_n(R)$. For example, if $n = 3$ then

$$v_1 = \left(\frac{1}{\sqrt{2}}, -\frac{1}{\sqrt{2}}, 0\right), \quad v_2 = \left(\frac{1}{\sqrt{2}}, \frac{1}{\sqrt{2}}, 0\right), \quad v_3 = (0,0,1)$$

is an o.n. basis of $V_3(R)$. The question of the existence of an o.n. basis with respect to a given inner product is settled in the next important theorem.

Before proceeding, we remark that if $u_1, \ldots, u_p$ are o.n. then they must be linearly independent. For, if

$$\sum_{j=1}^{p} c_j u_j = 0$$

then

$$0 = \left(\sum_{j=1}^{p} c_j u_j, u_k\right) = \sum_{j=1}^{p} c_j (u_j,u_k)$$
$$= \sum_{j=1}^{p} c_j \delta_{jk} = c_k, \quad k = 1, \ldots, p.$$

Theorem 3.2   (*Gram-Schmidt process*)   *Let V be a unitary space and let $u_1, \ldots, u_n$ be linearly independent vectors. Then there exists an* o.n. *set of vectors $v_1, \ldots, v_n$ such that*

$$\langle v_1, \ldots, v_k \rangle = \langle u_1, \ldots, u_k \rangle, \quad k = 1, \ldots, n. \tag{6}$$

Moreover, if $w_1, \ldots, w_k$ is any other o.n. *set such that*

$$\langle w_1, \ldots, w_k \rangle = \langle u_1, \ldots, u_k \rangle, \quad k = 1, \ldots, n,$$

then

$$w_k = c_k v_k, \quad k = 1, \ldots, n,$$

and

$$|c_k| = 1, \quad k = 1, \ldots, n.$$

Proof.   We prove the theorem by actually giving an algorithm for constructing the o.n. set $v_1, \ldots, v_n$. The vector u_1 is not 0 and hence $\|u_1\| = (u_1, u_1)^{1/2} \neq 0$. Set

$$v_1 = \frac{u_1}{\|u_1\|}$$

so that

$$\|v_1\| = \left\| \frac{u_1}{\|u_1\|} \right\| = \frac{1}{\|u_1\|} (u_1, u_1)^{1/2} = \frac{\|u_1\|}{\|u_1\|} = 1.$$

We have constructed the first vector in the sequence. Suppose $v_1, \ldots, v_p$ are already on hand and satisfy (6) for $k = 1, \ldots, p$. Our problem is to construct v_{p+1}. We start with the vector g_{p+1} defined by

$$g_{p+1} = u_{p+1} - \sum_{j=1}^{p} (u_{p+1}, v_j) v_j. \tag{7}$$

We first observe that $g_{p+1} \neq 0$. For if $g_{p+1} = 0$ then from (6)

$$u_{p+1} \in \langle v_1, \ldots, v_p \rangle = \langle u_1, \ldots, u_p \rangle.$$

But this implies that $u_1, \ldots, u_{p+1}$ are linearly dependent, a contradiction. Hence $g_{p+1} \neq 0$ and we can define v_{p+1} by

$$v_{p+1} = \frac{g_{p+1}}{\|g_{p+1}\|}.$$

We must prove two things: $v_1, \ldots, v_p, v_{p+1}$ are o.n. and (6) holds for $k = p + 1$. To check the first assertion observe that for $k \leq p$

$$(v_{p+1}, v_k) = \frac{1}{\|g_{p+1}\|} (g_{p+1}, v_k)$$

$$= \frac{1}{\|g_{p+1}\|} \left(u_{p+1} - \sum_{j=1}^{p} (u_{p+1}, v_j) v_j, v_k \right)$$

$$= \frac{1}{\|g_{p+1}\|} \left\{ (u_{p+1}, v_k) - \sum_{j=1}^{p} (u_{p+1}, v_j)(v_j, v_k) \right\}.$$

The vectors $v_1, \ldots, v_p$ are o.n. and hence for $k \leq p$, $(v_j, v_k) = \delta_{jk}$. We have

$$(v_{p+1}, v_k) = \frac{1}{\|g_{p+1}\|} \left\{ (u_{p+1}, v_k) - (u_{p+1}, v_k) \right\}$$
$$= 0.$$

From (7) and the definition of v_{p+1} we know that $u_{p+1} \in \langle v_1, \ldots, v_p, v_{p+1} \rangle$ and, since we have assumed that

$$\langle u_1, \ldots, u_p \rangle = \langle v_1, \ldots, v_p \rangle,$$

we can conclude that

$$\langle u_1, \ldots, u_p, u_{p+1} \rangle \subset \langle v_1, \ldots, v_{p+1} \rangle.$$

But $v_1, \ldots, v_{p+1}$ are o.n. and hence linearly independent. It follows that

$$\dim \langle v_1, \ldots, v_{p+1} \rangle = p + 1 = \dim \langle u_1, \ldots, u_{p+1} \rangle$$

and hence (6) holds for $k = p + 1$.

To prove the second part of the theorem, suppose $w_1, \ldots, w_n$ is an o.n. set and

$$\langle w_1, \ldots, w_k \rangle = \langle u_1, \ldots, u_k \rangle$$
$$= \langle v_1, \ldots, v_k \rangle, \quad k = 1, \ldots, n.$$

Then

$$w_p = \sum_{j=1}^{p} d_{pj} v_j$$

for appropriate complex numbers $d_{pj}, j = 1, \ldots, p$. But if $k \le p$, then

$$(w_p, v_k) = \sum_{j=1}^{p} d_{pj}(v_j, v_k) = d_{pk}.$$

However, if $k < p$ then $w_p \in \langle w_1, \ldots, w_k \rangle^{\perp} = \langle v_1, \ldots, v_k \rangle^{\perp}$, and thus $(w_p, v_k) = 0$. We conclude that $d_{pk} = 0$ for $k < p$. Set $d_{pp} = c_p$ and then

$$1 = (w_p, w_p) = (c_p v_p, c_p v_p) = |c_p|^2 (v_p, v_p) = |c_p|^2$$

which completes the proof of the theorem.

As an example of the preceding theorem, we find an orthonormalizing sequence v_1, v_2, v_3 for the following set of vectors in $V_4(\text{Re})$:

$$u_1 = (1, -1, 1, -1), \quad u_2 = (5, 1, 1, 1), \quad u_3 = (2, 3, 4, -1).$$

First,

$$v_1 = u_1 / \|u_1\| = \frac{1}{2}(1, -1, 1, -1).$$

Then,

$$g_2 = u_2 - (u_2, v_1) v_1$$
$$= (5, 1, 1, 1) - 2 \cdot \frac{1}{2}(1, -1, 1, -1)$$
$$= (4, 2, 0, 2),$$

and therefore

$$v_2 = g_2/\|g_2\| = \frac{1}{\sqrt{6}}(2,1,0,1).$$

Finally,

$$
\begin{aligned}
g_3 &= u_3 - (u_3,v_2)v_2 - (u_3,v_1)v_1 \\
&= (2,3,4,-1) - \frac{6}{\sqrt{6}} \cdot \frac{1}{\sqrt{6}}(2,1,0,1) - 2 \cdot \frac{1}{2}(1,-1,1,-1) \\
&= (-1,3,3,-1),
\end{aligned}
$$

and thus

$$v_3 = g_3/\|g_3\| = \frac{1}{\sqrt{20}}(-1,3,3,-1).$$

Theorem 3.2 allows us to prove our next result which has a very appealing geometric interpretation. Before doing this we define the idea of direct sum.

Definition 3.4 (*Sum of subspaces*) *If V is a vector space over an arbitrary field R and W_1 and W_2 are subspaces of V, then the **sum** of W_1 and W_2, denoted by $W_1 + W_2$, is the totality of vectors of the form $w_1 + w_2$, $w_1 \in W_1$, $w_2 \in W_2$. More generally, if $W_1, \ldots, W_p, \ldots$ is a sequence (not necessarily finite) of subspaces of V such that any vector in V can be expressed as a sum of a finite number of vectors chosen from $W_1, \ldots, W_p, \ldots$, then V is the sum of $W_1, \ldots, W_p, \ldots$, and is denoted by*

$$V = \sum_i W_i.$$

*If V is the sum of $W_1, \ldots, W_p, \ldots$ and if each vector in V can be expressed in only one way as a sum of a finite number of vectors, each from a different subspace W_i, then V is the **direct sum** of the subspaces W_i. This is denoted by*

$$V = W_1 \dotplus W_2 \dotplus \cdots \dotplus W_p \dotplus \cdots$$

or

$$V = \sum_i{}^{\cdot} W_i.$$

Notice that in this definition we have insisted that the vectors in V are finite sums of vectors from the subspaces W_i. This relieves us of the whole problem of convergence of sequences of vectors and is entirely satisfactory for most algebraic situations. It is clear that the sum of subspaces is always a subspace. Moreover the set intersection of any collection of subspaces is also always a subspace (i.e., a linear combination of vectors in every one of the subspaces of the collection is obviously in every one of the subspaces as well). We shall use the familiar notation for the intersection of subspaces: if W_1 and W_2 are subspaces then $W_1 \cap W_2$ is the set of all vectors belonging to both W_1 and W_2. More generally, if $W_1, W_2, \ldots$ is any set of subspaces then $\underset{i}{\cap}\, W_i$ is the set of all vectors lying in every W_i.

We remark that we could have defined the direct sum of two subspaces W_1 and W_2 of V as follows:

$$V = W_1 \dotplus W_2 \quad \text{if } V = W_1 + W_2 \text{ and } W_1 \cap W_2 = 0.$$

For, if $v = w_1 + w_2 = w_1' + w_2'$, where $w_1, w_1' \in W_1$ and $w_2, w_2' \in W_2$, then $w_1 - w_1' = w_2' - w_2$. But $w_1 - w_1' \in W_1$ and $w_2' - w_2 \in W_2$ and since $W_1 \cap W_2 = 0$ we conclude that $w_1 = w_1'$ and $w_2 = w_2'$.

Theorem 3.3 *If $W_1, \ldots, W_p$ are finite dimensional subspaces of a vector space V over an arbitrary field R then*

$$\dim \left(\sum_{i=1}^{p} W_i \right) \leq \sum_{i=1}^{p} \dim W_i. \tag{8}$$

In general,

$$\dim (W_1 + W_2) = \dim W_1 + \dim W_2 - \dim (W_1 \cap W_2). \tag{9}$$

If R is either the real or complex field, and V is finite dimensional and W is a subspace of V then

$$V = W \dotplus W^{\perp}. \tag{10}$$

Equation (10) is known as the "projection theorem."

Proof. The proof of the inequality (8) is very easy. Let

$$w_{i,1}, \ldots, w_{i,n_i} \tag{11}$$

be a basis of W_i, $i = 1, \ldots, p$. Then any vector $w \in \sum\limits_{i=1}^{p} W_i$ is a sum of vectors $w = w_1 + \cdots + w_p$, $w_i \in W_i$, $i = 1, \ldots, p$. Now each w_i is a linear combination of the vectors (11) and hence w itself is a linear combination of the totality of $n_1 + \cdots + n_p$ vectors appearing among all the sets (11). Thus

$$\dim \left(\sum_{i=1}^{p} W_i \right) \le n_1 + \cdots + n_p = \sum_{i=1}^{p} \dim W_i.$$

The proof of (9) is a little harder. We begin by obtaining a basis $u_1, \ldots, u_r$ of the subspace $W_1 \cap W_2$. By Theorem 1.1 (iii), we can complete $u_1, \ldots, u_r$ first to a basis of W_1 and then to a basis of W_2. In other words, let

$$u_1, \ldots, u_r, w_{r+1}, \ldots, w_{n_1}$$

be a basis of W_1 and

$$u_1, \ldots, u_r, v_{r+1}, \ldots, v_{n_2}$$

be a basis of W_2. Any vector in $W_1 + W_2$ is clearly a linear combination of the vectors $u_1, \ldots, u_r, w_{r+1}, \ldots, w_{n_1}, v_{r+1}, \ldots, v_{n_2}$. We prove that these vectors are linearly independent. Let $a_1, \ldots, a_r$, $b_{r+1}, \ldots, b_{n_1}, c_{r+1}, \ldots, c_{n_2}$ be any scalars for which

$$\sum_{i=1}^{r} a_i u_i + \sum_{i=r+1}^{n_1} b_i w_i + \sum_{i=r+1}^{n_2} c_i v_i = 0. \tag{12}$$

We want to show that all these scalars must be 0. To do this let u, w, and v denote the three sums in (12) respectively so that

$$u + w + v = 0. \tag{13}$$

Now (13) implies that $w = -u - v \in W_2$ and since $w \in W_1$ we have

$$w \in W_1 \cap W_2 = \langle u_1, \ldots, u_r \rangle.$$

But

$$w = \sum_{i=r+1}^{n_1} b_i w_i$$

and the vectors $u_1, \ldots, u_r, w_{r+1}, \ldots, w_{n_1}$ form a basis of W_1. We have proved that w is a linear combination of $u_1, \ldots, u_r$ and at the same time a linear combination of $w_{r+1}, \ldots, w_{n_1}$. There is only one possibility: $w = 0$. Hence $b_{r+1} = \cdots = b_{n_1} = 0$. But $u + v = 0$ implies a linear dependence relation among the basis vectors $u_1, \ldots, u_r$, $v_{r+1}, \ldots, v_{n_2}$ unless all the coefficients a_i and c_i are zero. We have proved that the vectors $u_1, \ldots, u_r, w_{r+1}, \ldots, w_{n_1}, v_{r+1}, \ldots, v_{n_2}$ constitute a basis for $W_1 + W_2$. Hence

$$\begin{aligned}
\dim (W_1 + W_2) &= r + (n_1 - r) + (n_2 - r) \\
&= n_1 + n_2 - r \\
&= \dim W_1 + \dim W_2 - \dim (W_1 \cap W_2).
\end{aligned}$$

To establish (10) we can argue as follows. Let $u_1, \ldots, u_m$ form a basis of W, and augment these with $u_{m+1}, \ldots, u_n$ to a basis of V. By Theorem 3.2, we can obtain an o.n. basis $v_1, \ldots, v_n$ of V such that

$$\langle v_1, \ldots, v_m \rangle = \langle u_1, \ldots, u_m \rangle = W.$$

If $v \in \langle v_{m+1}, \ldots, v_n \rangle$ and $w \in W$ then the orthogonality of the vectors v_i clearly implies that $(v,w) = 0$. Hence

$$\langle v_{m+1}, \ldots, v_n \rangle \subset W^\perp. \tag{14}$$

Since $\langle v_1, \ldots, v_m \rangle = W$ and $v_1, \ldots, v_n$ is a basis of V we see from (14) that $V = W + W^\perp$. Now $W \cap W^\perp = \{0\}$. To see this suppose that x is a vector in both W and $W^\perp$. Then $(x,x) = 0$ and it follows that $x = 0$. From (9) we have

$$\begin{aligned}
n = \dim V = \dim (W + W^\perp) &= \dim W + \dim W^\perp \\
&= m + \dim W^\perp.
\end{aligned}$$

Thus $\dim W^\perp = n - m$. It follows from Theorem 1.3 that the inclusion (14) is actually equality.

An interesting geometrical problem can be solved (at least in principle!) by use of Theorem 3.2. The problem is this: given a finite dimensional space V, a subspace W with the basis $u_1, \ldots, u_k$ and a vector u not in W, find the distance from u to W, i.e., find the minimum length of the difference vectors $u - x$ as x varies over W. This is, of course, a generalization of the familiar problem in analytic geometry of finding the distance from a point to a plane. The distance from a point (y_1,y_2,y_3)

to a plane is in fact the minimum of the distances between (y_1, y_2, y_3) and points (x_1, x_2, x_3) lying in the plane. Let $u_{k+1}, \ldots, u_n$ be a set of vectors such that $u_1, \ldots, u_n$ is a basis of V. Let $v_1, \ldots, v_n$ be an o.n. basis of V satisfying

$$\langle v_1, \ldots, v_p \rangle = \langle u_1, \ldots, u_p \rangle, \quad p = 1, \ldots, n.$$

Thus, if u is the given vector and x is any vector in W then

$$u = \sum_{i=1}^{n} c_i v_i \quad \text{and} \quad x = \sum_{i=1}^{k} x_i v_i$$

for appropriate complex numbers c_i and x_i. Let $d(x)$ denote the length of the vector $u - x$ and then we have

$$
\begin{aligned}
d^2(x) &= (u - x, u - x) \\
&= \Big(\sum_{i=1}^{n} c_i v_i - \sum_{i=1}^{k} x_i v_i, \ \sum_{i=1}^{n} c_i v_i - \sum_{i=1}^{k} x_i v_i \Big) \\
&= \sum_{i=1}^{k} |c_i - x_i|^2 + \sum_{i=k+1}^{n} |c_i|^2.
\end{aligned}
$$

It is clear that $d(x)$ will be minimum when $x_i = c_i$, $i = 1, \ldots, k$. For this choice of x the square of the distance is then given by

$$\sum_{i=k+1}^{n} |c_i|^2.$$

In the next theorem we collect together several classical results of great importance. It has been said that the celebrated Cauchy-Schwarz inequality in Theorem 3.4 is the "only inequality" in algebra, i.e., that every other inequality can be deduced from it. The problem is to find a suitable unitary space!

Theorem 3.4 *Let V be a unitary space with inner product (x, y).*

(a) *(Bessel's inequality) If $v_1, \ldots, v_n$ is an o.n. set in V and $x \in V$, then*

$$\|x\|^2 \geq \sum_{j=1}^{n} |(x, v_j)|^2. \tag{15}$$

Equality holds in (15) if and only if $x \in \langle v_1, \ldots, v_n \rangle$.

(b)   *(Parseval's identity)   If $v_1, \ldots, v_n$ is an o.n. set and x and y are in $\langle v_1, \ldots, v_n \rangle$ then*

$$(x,y) = \sum_{j=1}^{n} (x,v_j)(v_j,y). \tag{16}$$

(c)   *(Cauchy-Schwarz inequality)   If x and y are any two vectors in V, then*

$$|(x,y)| \leq \|x\| \, \|y\|. \tag{17}$$

Equality holds in (17) if and only if x and y are linearly dependent.

(d)   *(Triangle inequality)   If x and y are in V then*

$$\|x + y\| \leq \|x\| + \|y\|. \tag{18}$$

Equality holds in (18) if and only if one of the vectors is zero or neither is zero and $y = cx$, $c > 0$.

Proof.   (a)   We know that

$$\left\| x - \sum_{j=1}^{n} (x,v_j)v_j \right\|^2 \geq 0 \tag{19}$$

with equality holding in (19) if and only if

$$x = \sum_{j=1}^{n} (x,v_j)v_j. \tag{20}$$

But

$$0 \leq \left\| x - \sum_{j=1}^{n} (x,v_j)v_j \right\|^2$$

$$= \left(x - \sum_{j=1}^{n} (x,v_j)v_j, \; x - \sum_{j=1}^{n} (x,v_j)v_j \right)$$

$$= (x,x) - \sum_{j=1}^{n} (\overline{x,v_j})(x,v_j) - \sum_{j=1}^{n} (x,v_j)(v_j,x) + \sum_{j,k=1}^{n} (x,v_j)(\overline{x,v_k})(v_j,v_k)$$

$$= \|x\|^2 - \sum_{j=1}^{n} |(x,v_j)|^2 - \sum_{j=1}^{n} (x,v_j)(\overline{x,v_j}) + \sum_{j,k=1}^{n} (x,v_j)(\overline{x,v_k})\delta_{jk}$$

$$= \|x\|^2 - 2 \sum_{j=1}^{n} |(x,v_j)|^2 + \sum_{j=1}^{n} |(x,v_j)|^2$$

$$= \|x\|^2 - \sum_{j=1}^{n} |(x,v_j)|^2.$$

(b)   We can write $x = \sum_{j=1}^{n} c_j v_j$ and then immediately compute that $(x,v_k) = c_k$. From this we see that

$$(x,y) = \Big(\sum_{j=1}^{n} (x,v_j)v_j, \sum_{k=1}^{n} (y,v_k)v_k \Big)$$

$$= \sum_{j,k=1}^{n} (x,v_j)(\overline{y,v_k})(v_j,v_k)$$

$$= \sum_{j,k=1}^{n} (x,v_j)(\overline{y,v_k})\delta_{jk}$$

$$= \sum_{j=1}^{n} (x,v_j)(v_j,y).$$

(c)   If $x = 0$ then both sides of (17) are zero: x and y are then linearly dependent. Assume then that $x \neq 0$ and define

$$w = y - \Big(y, \frac{x}{\|x\|}\Big) \frac{x}{\|x\|},$$

i.e., subtract from y its projection in the direction of x. Then

$$(w,x) = (y,x) - \frac{(y,x)}{\|x\|^2}(x,x) = (y,x) - (y,x) = 0.$$

Hence,

$$\|w\|^2 = (w,w) = \Big(w, y - \frac{(y,x)}{\|x\|^2}x\Big)$$

$$= (w,y) - \frac{(\overline{y,x})}{\|x\|^2}(w,x)$$

$$= (w,y)$$

$$= \Big(y - \frac{(y,x)}{\|x\|^2}x, y\Big)$$

$$= \|y\|^2 - \frac{(y,x)(x,y)}{\|x\|^2}.$$

Now $\|w\|^2 \geq 0$ and thus

$$\|y\|^2\|x\|^2 \geq (y,x)(x,y) = |(x,y)|^2.$$

Moreover, if equality holds then $\|w\| = 0$, i.e., $w = 0$. But this means that x and y are linearly dependent. Conversely, if x and y are linearly dependent and $x \neq 0$ then $y = cx$ and

$$|(x,y)| = |c|\,\|x\|^2,$$

while

$$\|x\|\,\|y\| = \|x\|\,\|cx\| = |c|\,\|x\|^2,$$

and equality holds in (17).

(d) We have

$$\begin{aligned}
\|x + y\|^2 &= (x + y, x + y) = (x,x) + (y,y) + (x,y) + (y,x)\\
&= \|x\|^2 + \|y\|^2 + (x,y) + \overline{(x,y)} \qquad\qquad (21)\\
&\leq \|x\|^2 + \|y\|^2 + 2|(x,y)|,
\end{aligned}$$

since the real part of (x,y) cannot exceed $|(x,y)|$. Hence, by (17),

$$\begin{aligned}
\|x + y\|^2 &\leq \|x\|^2 + \|y\|^2 + 2\|x\|\,\|y\| \qquad\qquad (22)\\
&= (\|x\| + \|y\|)^2
\end{aligned}$$

and (18) follows. From the preceding calculation we see that if equality holds in (18) then equality must hold in both (21) and (22). Now equality in (21) implies

$$(x,y) + (y,x) = 2|(x,y)|. \qquad\qquad (23)$$

The equality in (22) can hold [according to (c)] if and only if x and y are linearly dependent. If x and y are not zero then $y = cx$ and from (23) we have

$$\begin{aligned}
(x,y) + (y,x) &= (x,cx) + (cx,x) = (c + \bar{c})\|x\|^2,\\
2|(x,y)| &= 2|(x,cx)| = 2|c|\,\|x\|^2,
\end{aligned}$$

so that $c + \bar{c} = 2|c|$ and it follows that c is real and positive. Conversely, if $y = cx$, $c > 0$ or if either vector is 0 then (18) is clearly equality.

The inequality (17) is really quite an extraordinary result. Let us apply it to the vector space $V_n(R)$, where R is the complex number field (using the standard inner product). Set

$$u = (u_1, \ldots, u_n), \quad v = (v_1, \ldots, v_n).$$

Then

$$(u,v) = \sum_{i=1}^{n} u_i \bar{v}_i$$

and

$$\|u\|^2 = \sum_{i=1}^{n} |u_i|^2, \quad \|v\|^2 = \sum_{i=1}^{n} |v_i|^2.$$

Thus (17) specializes to

$$\left| \sum_{i=1}^{n} u_i \bar{v}_i \right|^2 \le \sum_{i=1}^{n} |u_i|^2 \sum_{i=1}^{n} |v_i|^2 \tag{24}$$

with equality if and only if u and v are linearly dependent. For example, we can prove the following elementary inequality: *if $a_1, \ldots, a_n$ are positive numbers, then*

$$\sum_{i=1}^{n} a_i \sum_{i=1}^{n} a_i^{-1} \ge n^2 \tag{25}$$

with equality if and only if $a_1 = \cdots = a_n$. For, let

$$u = (\sqrt{a_1}, \ldots, \sqrt{a_n}) \quad \text{and} \quad v = \left(\frac{1}{\sqrt{a_1}}, \ldots, \frac{1}{\sqrt{a_n}} \right).$$

Then directly from (24) we compute

$$n = \sum_{i=1}^{n} \sqrt{a_i} \frac{1}{\sqrt{a_i}} = |(u,v)| \le \|u\| \, \|v\|$$

$$= \left(\sum_{i=1}^{n} a_i \right)^{1/2} \left(\sum_{i=1}^{n} a_i^{-1} \right)^{1/2}.$$

Equality holds if and only if $u = cv$ for some positive scalar c. But then $a_i = c^2 a_i^{-1}$, $a_i^2 = c^2$ and hence all the a_i are equal.

Quiz

Answer true or false:

1. Any vector in a unitary space is perpendicular to the zero vector.
2. Let u and v be two linearly independent vectors in a finite dimensional vector space V over the complex numbers. Then there exists an inner product for V in which $(u,v) = 0$, $\|u\|^2 = \|v\|^2 = 1$.
3. If V is a unitary space and $x \in \langle y, z\rangle^\perp$ then x, y, and z are linearly independent.
4. If $V = V_3(R)$, R the real number field, then the set of vectors perpendicular to the plane $x_1 - 2x_2 + 3x_3 = 0$ span a one dimensional subspace of V. (Use the standard inner product.)
5. If $v \in V$, and V is a finite dimensional unitary space then $(\langle v\rangle^\perp)^\perp = \langle v\rangle$.
6. If $v \in V$, V is a unitary space, dim $V = n$, and v is perpendicular to n linearly independent vectors then $v = 0$.
7. If W_1 and W_2 are subspaces of a finite dimensional space V, dim $V = n$, then dim W_1 dim $W_2 \leq n^2/4$.
8. Under the same conditions as in the preceding problem: if dim $W_1 > n/2$ and dim $W_2 > n/2$ then dim $(W_1 \cap W_2) \geq 1$.
9. If W_1 and W_2 are the subspaces in Question 7 above, V is a unitary space with a fixed inner product, and $W_1 \cap W_2 = 0$ then bases $u_1, \ldots, u_p$ and $v_1, \ldots, v_r$ may be chosen for W_1 and W_2 respectively such that $(u_i,v_j) = 0$ for all $i = 1, \ldots, p, j = 1, \ldots, r$.
10. Let $(x,y) = \sum_{i=1}^{n} x_i y_i$ be defined on pairs of vectors $x = (x_1, \ldots, x_n)$ and $y = (y_1, \ldots, y_n)$ in $V_n(R)$, R the complex field. Then (x,y) is an inner product.

Exercises

1. Let $V = V_3(R)$ where R is the field of complex numbers. Let $v_1 = (1,0,-1)$, $v_2 = (0,0,1)$, $v_3 = (0,i,-1)$. If $x = (2,3,0)$ and $y = (1,-1,1)$ compute the value of the standard inner product (x,y) relative to the basis v_1,v_2,v_3.
2. Verify Def. 3.1 (i), (ii) for the standard inner product (see the proof of Theorem 3.1).
3. If V is a unitary space and $v_1, \ldots, v_n$ are nonzero vectors in V satisfying $(v_i,v_j) = 0$, $i \neq j$, show that they are linearly independent.
4. Let $a_1, \ldots, a_n$ be positive numbers and suppose λ is any real number.

Show that

$$\left(\sum_{i=1}^{n} a_i^{1/2}\right)^2 \le \sum_{i=1}^{n} a_i^{\lambda} \sum_{i=1}^{n} a_i^{1-\lambda}$$

with equality if and only if either $\lambda = \frac{1}{2}$ or $\lambda \ne \frac{1}{2}$ and all the a_i are equal.

5. Let V be a finite dimensional unitary space. If (u,v) and $(u,v)_1$ are two inner products show that there exists a constant c independent of u such that $(u,u)_1 \le c(u,u)$ for all $u \in V$.

6. If V is a finite dimensional unitary space then show that the distance from a vector v to its orthogonal complement $\langle v \rangle^{\perp}$ is equal to $\|v\|$.

7. Using the method in the text, find the distance from the point $(1,3,1)$ to the plane through the origin and containing the points $(1,2,2)$ and $(3,1,2)$. The vector space is $V_3(R)$, R the real number field, with the standard inner product.

8. In $V_n(R)$, R the complex field, show that

$$(u,v) = \sum_{i=1}^{n} \lambda_i u_i \bar{v}_i, \quad u = (u_1, \ldots, u_n), \quad v = (v_1, \ldots, v_n)$$

is an inner product if $\lambda_i > 0$, $i = 1, \ldots, n$.

9. Let V be a finite dimensional unitary space with an o.n. basis $v_1, \ldots, v_n$. Show that if S and T are in $L(V,V)$ and (S,T) is defined by

$$(S,T) = \sum_{i=1}^{n} (Sv_i, Tv_i)$$

then $L(V,V)$ together with the function (S,T) is a unitary space. In other words, show that (S,T) satisfies the definition of an inner product.

10. Find an o.n. basis for the inner product in $L(V,V)$ given in Exercise 9.

1.4 Matrix Representations

Suppose that U and V are finite dimensional vector spaces over a field R and let $T \in L(U,V)$. As we know from our discussion of the process of linear extension in Theorem 2.1, the transformation T is completely determined by its values on a basis of U. It is perhaps worthwhile to go over this point again. Let $G = \{g_1, \ldots, g_n\}$ be a basis of U and suppose

$$T(g_j) = v_j, \quad j = 1, \ldots, n. \tag{1}$$

Thus if u is any vector in U,

$$u = \sum_{j=1}^{n} c_j g_j,$$

and we have from (1)

$$T(u) = T\left(\sum_{j=1}^{n} c_j g_j\right)$$

$$= \sum_{j=1}^{n} c_j T(g_j)$$

$$= \sum_{j=1}^{n} c_j v_j.$$

Hence $T(u)$ is completely determined by the values $T(g_j) = v_j$, $j = 1$, . . . , n.

Suppose next that $H = \{h_1, \ldots, h_m\}$ is a basis of V. Then each vector v_j in (1) has a unique representation as a linear combination of $h_1, \ldots, h_m$:

$$T(g_j) = v_j = \sum_{i=1}^{m} a_{ij} h_i, \quad j = 1, \ldots, n. \tag{2}$$

Thus the transformation T is known completely if we know the mn numbers a_{ij}, $i = 1, \ldots, m$, $j = 1, \ldots, n$, and the bases G and H.

Definition 4.1 (**Matrices**) *If a_{ij}, $i = 1, \ldots, m$, $j = 1, \ldots, n$, are mn numbers in a ring R then the rectangular array*

$$A = \begin{bmatrix} a_{11} & a_{12} & a_{13} & \cdots & a_{1n} \\ a_{21} & a_{22} & a_{23} & \cdots & a_{2n} \\ \cdot & & & & \\ \cdot & & & & \\ \cdot & & & & \\ a_{m1} & a_{m2} & a_{m3} & \cdots & a_{mn} \end{bmatrix} \tag{3}$$

*is called an $m \times n$ **matrix** over R. If A is the matrix defined by the linear transformation $T \in L(U,V)$ and the pair of bases G and H as in (2) then we say that A is the **matrix representation** of T with respect to the bases G and H. This is written*

$$A = [T]_G{}^H. \tag{4}$$

According to the remarks immediately preceding Def. 4.1, *there is one and only one matrix representation for $T \in L(U,V)$, once the bases G and H are fixed.* For, the elements of A tell us how to express each vector $T(g_j)$ in terms of the basis vectors H and then T is determined by linear extension from the values $T(g_j)$.

There are many special notations and definitions having to do with matrices. We list some of these.

Definition 4.2 (*Notation for matrices*) *The following pertain to the matrix A in (3).*

(a) *The scalar a_{ij} is called the (i,j) **entry** or **element** of A. It is sometimes denoted by A_{ij} as well. Thus we sometimes write $A = (a_{ij})$.*

(b) *The i-th **row** of A is the n-tuple $(a_{i1}, a_{i2}, \ldots, a_{in})$. It is designated by $A_{(i)}$.*

(c) *The j-th **column** of A is the m-tuple $(a_{1j}, a_{2j}, \ldots, a_{mj})$ and is designated by $A^{(j)}$.*

(d) *The totality of $m \times n$ matrices (i.e., m rows and n columns) with entries in R will be designated by $M_{m,n}(R)$. If $m = n$ we say A is n-**square** and the totality of n-square matrices is designated by $M_n(R)$.*

(e) *The matrix in $M_{m,n}(R)$ with every entry equal to 0 is called the **zero** matrix and is denoted by $0_{m,n}$ or just 0 when m and n are understood.*

(f) *The matrix in $M_n(R)$ whose (i,j) entry is δ_{ij}, $i, j = 1, \ldots, n$, is called the **identity** matrix and is denoted by I_n.*

(g) *If $\alpha_1, \ldots, \alpha_n$ are elements of R then the matrix A whose (i,j) entry is $\alpha_i \delta_{ij}$ is called a **diagonal** matrix. It is written*

$$A = \text{diag} \, (\alpha_1, \ldots, \alpha_n).$$

(h) *The **main diagonal** of a matrix $A \in M_n(R)$ is the n-tuple of elements lying in positions (i,i), $i = 1, \ldots, n$.*

There are various ways of combining matrices but before we formally define these we shall prove a theorem that will motivate the definitions.

Theorem 4.1 *Let U,V,W be vector spaces over a field with bases $G = \{g_1, \ldots, g_n\}$, $H = \{h_1, \ldots, h_m\}$ and $K = \{k_1, \ldots, k_r\}$ respectively. Let $T \in L(U,V)$, $S \in L(V,W)$ and let P denote the product $P = ST \in L(U,W)$.*

Define

$$A = (a_{ij}) = [T]_G{}^H \in M_{m,n}(R),$$
$$B = (b_{ij}) = [S]_H{}^K \in M_{r,m}(R)$$

and

$$C = (c_{ij}) = [P]_G{}^K \in M_{r,n}(R).$$

Then

$$c_{ij} = \sum_{s=1}^{m} b_{is}a_{sj}, \quad i = 1, \ldots, r, \quad j = 1, \ldots, n. \tag{5}$$

If $\alpha \in R$, then

$$[\alpha T]_G{}^H = (\alpha a_{ij}). \tag{6}$$

If $M \in L(U,V)$ and $[M]_G{}^H = D$ then

$$[T + M]_G{}^H = (a_{ij} + d_{ij}). \tag{7}$$

Proof. To prove (5) we must compute $[ST]_G{}^K$. Thus we must examine STg_j:

$$STg_j = S(Tg_j) \tag{8}$$

$$= S\left(\sum_{s=1}^{m} a_{sj}h_s\right)$$

$$= \sum_{s=1}^{m} a_{sj}S(h_s)$$

$$= \sum_{s=1}^{m} a_{sj} \sum_{i=1}^{r} b_{is}k_i$$

$$= \sum_{i=1}^{r} \left(\sum_{s=1}^{m} b_{is}a_{sj}\right) k_i.$$

On the other hand, from the definition of C we have

$$(ST)g_j = \sum_{i=1}^{r} c_{ij}k_i. \tag{9}$$

The k_i are linearly independent. Hence matching coefficients of the k_i in (8) and (9) we have

$$c_{ij} = \sum_{s=1}^{m} b_{is}a_{sj}, \quad i = 1, \ldots, r, \quad j = 1, \ldots, n.$$

To obtain (6) we note that

$$(\alpha T)g_j = \alpha T(g_j) = \alpha \sum_{i=1}^{m} a_{ij}h_i$$
$$= \sum_{i=1}^{m} (\alpha a_{ij})h_i.$$

Thus the coefficient of h_i in the expression for $(\alpha T)g_j$ is αa_{ij}. This establishes (6).

The proof of (7) is quite similar:

$$(T + M)g_j = Tg_j + Mg_j$$
$$= \sum_{i=1}^{m} a_{ij}h_i + \sum_{i=1}^{m} d_{ij}h_i$$
$$= \sum_{i=1}^{m} (a_{ij} + d_{ij})h_i.$$

Hence (7) follows.

We are thus led to the following definitions for operations between matrices.

Definition 4.3 *(Matrix operations)* Let R be a commutative ring. Let $A \in M_{m,n}(R)$, $B \in M_{r,m}(R)$, and $D \in M_{m,n}(R)$. Let $\alpha \in R$.
  (a) The **sum** of A and D is the $m \times n$ matrix whose (i,j) entry is $a_{ij} + d_{ij}$, $i = 1, \ldots, m$, $j = 1, \ldots, n$. The sum is designated by $A + D$. Thus $A + 0_{m,n} = 0_{m,n} + A = A$.
  (b) The **product** of B and A is the $r \times n$ matrix C whose (i,j) entry is

$$c_{ij} = \sum_{s=1}^{m} b_{is}a_{sj}, \quad i = 1, \ldots, r, \quad j = 1, \ldots, n. \quad (10)$$

Thus $AI_n = A = I_mA$. The product is denoted by juxtaposing the two matrices:

$$C = BA.$$

(c) *The **scalar product** of α and A is the matrix whose (i,j) entry is αa_{ij}, $i = 1, \ldots, m$, $j = 1, \ldots, n$. The scalar product is denoted by αA or $A\alpha$.*

The first thing we observe is that the Def. 4.3 is so constructed that in the notation of Theorem 4.1 we can immediately conclude

$$[T + M]_G{}^H = [T]_G{}^H + [M]_G{}^H, \tag{11}$$
$$[ST]_G{}^K = [S]_H{}^K[T]_G{}^H, \tag{12}$$
$$[\alpha T]_G{}^H = \alpha[T]_G{}^H. \tag{13}$$

The formulas (11) and (12) imply the associative and distributive laws for matrix multiplication as defined in Def. 4.3. To be specific about this, suppose $A \in M_{m,n}(R)$, $B \in M_{r,m}(R)$, $E \in M_{p,r}(R)$. Let U, V, W, and X be vector spaces over R with bases G, H, K, N respectively. Let $T_1: U \to V$, $T_2: V \to W$, $T_3: W \to X$, be chosen so that $[T_1]_G{}^H = A$, $[T_2]_H{}^K = B$ and $[T_3]_K{}^N = E$. Then

$$\begin{aligned} E(BA) &= [T_3]_K{}^N[T_2T_1]_G{}^K = [T_3(T_2T_1)]_G{}^N = [(T_3T_2)T_1]_G{}^N \\ &= [T_3T_2]_H{}^N[T_1]_G{}^H = ([T_3]_K{}^N[T_2]_H{}^K)[T_1]_G{}^H \\ &= (EB)A. \end{aligned}$$

In other words, *matrix multiplication is associative*. Similarly the *distributive* laws hold for matrix multiplication. If $A \in M_n(R)$ and k is a nonnegative integer we can unambiguously define the matrix $A^k \in M_n(R)$, e.g., $A^3 = A^2A = AA^2$, $A^4 = A^3A = AA^3$, etc. We will define $A^0 = I_n$ for convenience.

As an example of computing the matrix representation of a linear transformation, let $U = V_3(R)$, $V = V_2(R)$ where R is the field of real numbers. Let $T \in L(U,V)$ be defined by

$$T(x_1, x_2, x_3) = (7x_1 + 2x_2 - 3x_3, x_2). \tag{14}$$

As bases for U and V respectively let

$$G = \{g_1, g_2, g_3\} = \{(1,0,0,), (0,1,-1), (0,0,1)\},$$
$$H = \{h_1, h_2\} = \{(1,0), (0,-1)\}.$$

Using (14) we compute

$$T(g_1) = T(1,0,0) = (7,0) = 7h_1 + 0h_2$$
$$T(g_2) = T(0,1,-1) = (5,1) = 5h_1 - 1h_2$$
$$T(g_3) = T(0,0,1) = (-3,0) = -3h_1 + 0h_2.$$

Hence

$$[T]_G{}^H = \begin{bmatrix} 7 & 5 & -3 \\ 0 & -1 & 0 \end{bmatrix}.$$

Suppose that U and V are finite dimensional unitary spaces. We should like to see what more can be said about matrix representations of linear transformations in such spaces.

Theorem 4.2 *Let U and V be finite dimensional unitary spaces and suppose that $T \in L(U,V)$. If $G = \{g_1, \ldots, g_n\}$ is a basis for U, $H = \{h_1, \ldots, h_m\}$ is an orthonormal basis for V, and $[T]_G{}^H = A = (a_{ij})$ then*

$$a_{ij} = (Tg_j, h_i), \quad i = 1, \ldots, m, \quad j = 1, \ldots, n. \tag{15}$$

Proof. By the definition of $[T]_G{}^H$ we have

$$Tg_j = \sum_{s=1}^{m} a_{sj}h_s.$$

Hence

$$(Tg_j, h_i) = \Big(\sum_{s=1}^{m} a_{sj}h_s, h_i \Big)$$
$$= \sum_{s=1}^{m} a_{sj}(h_s, h_i)$$
$$= a_{ij}.$$

Definition 4.4 (Transpose, conjugate) *If $A \in M_{m,n}(R)$ and R is any ring, then the **transpose** of A is the matrix in $M_{n,m}(R)$ whose (i,j) entry is the (j,i) entry of A. The transpose of A is denoted by A^T.*

Thus

$$(A^T)_{ij} = a_{ji}, \quad i = 1, \ldots, n, \quad j = 1, \ldots, m. \quad (16)$$

If R is the complex field then the **conjugate** *of A is the m × n matrix whose (i,j) entry is the complex conjugate of the (i,j) entry of A. The conjugate is denoted by $\bar{A}$ and thus*

$$(\bar{A})_{ij} = \bar{a}_{ij}, \quad i = 1, \ldots, m, \quad j = 1, \ldots, n. \quad (17)$$

The **conjugate transpose** *of A is the matrix $\bar{A}^T = \overline{A^T}$. The notation for the conjugate transpose is A^*. Thus*

$$(A^*)_{ij} = \bar{a}_{ji}, \quad i = 1, \ldots, n, \quad j = 1, \ldots, m. \quad (18)$$

There are several elementary rules for the conjugate transpose. The proofs of these rules will be enshrined in the exercises (see Exercise 3). Let A and B be $m \times n$ matrices over the complex numbers and let α be a complex number. Then

$$(A + B)^T = A^T + B^T, \quad \overline{A + B} = \bar{A} + \bar{B},$$
$$(A + B)^* = A^* + B^*. \quad (19)$$

If A is $m \times n$ and B is $n \times p$ then

$$(AB)^T = B^T A^T, \quad \overline{AB} = \bar{A}\,\bar{B}, \quad (AB)^* = B^*A^*; \quad (20)$$

$$(\alpha A)^T = \alpha A^T, \quad \overline{\alpha A} = \bar{\alpha}\,\bar{A}, \quad (\alpha A)^* = \bar{\alpha}\,A^*. \quad (21)$$

Let us return to the content of Theorem 4.2. There $T \in L(U,V)$ and $[T]_G{}^H = A \in M_{m,n}(R)$ where R is the field of complex numbers. Suppose that we look at the matrix $A^* \in M_{n,m}(R)$ and we want to define a linear transformation having A^* as its matrix representation.

Definition 4.5 **(Adjoint)** *Let U and V be unitary spaces and let $G = \{g_1, \ldots, g_n\}$ be an o.n. basis of U. Let $T \in L(U,V)$ and define a linear transformation $S: V \to U$ by*

$$Sv = \sum_{i=1}^{n} (v, Tg_i)g_i \quad (22)$$

*for each $v \in V$. The transformation S is called the **adjoint** of T and is denoted by T^*.*

The adjoint of T is defined in terms of the o.n. basis G. We shall see that the transformation T^* is in fact independent of any o.n. basis used to define it. However, it most certainly does depend on the inner product in V. That is, were we to change the inner product in V we would get a different transformation as the adjoint.

Another point to mention is the following: in talking about the two unitary spaces U and V, there is no reason to think that they should have the same inner product. However, we shall not distinguish between them notationally.

Theorem 4.3   *If $T \in L(U,V)$ as in Def. 4.5 then*

$$(Tu,v) = (u,T^*v) \tag{23}$$

for any $u \in U$ and $v \in V$. If $S_1 \in L(V,U)$ and

$$(Tu,v) = (u,S_1v)$$

for all $u \in U$, $v \in V$ then $S_1 = T^*$. If $G = \{g_1, \ldots, g_n\}$ and $H = \{h_1, \ldots, h_m\}$ are o.n. bases of U and V respectively, and $A = [T]_G{}^H$ then

$$A^* = [T^*]_H{}^G. \tag{24}$$

The transformation T^* does not depend on the basis used to define it in the following sense: if in (22) we were to use an o.n. basis $g_1', \ldots, g_n'$ and define S_1 by

$$S_1v = \sum_{i=1}^{n} (v,Tg_i')g_i', \tag{25}$$

then

$$S_1 = T^*.$$

Proof. To see (23) we compute that

$$(u, T^*v) = \left(u, \sum_{i=1}^{n} (v, Tg_i)g_i\right)$$

$$= \sum_{i=1}^{n} (u, g_i)\overline{(v, Tg_i)}$$

$$= \sum_{i=1}^{n} (u, g_i)(Tg_i, v)$$

$$= \left(\sum_{i=1}^{n} (u, g_i)Tg_i, v\right)$$

$$= \left(T\left[\sum_{i=1}^{n} (u, g_i)g_i\right], v\right)$$

$$= (Tu, v).$$

This proves (23). To prove (24) we use Theorem 4.2. Thus set $B = [T^*]_H{}^G$ and then

$$b_{ij} = (T^*h_j, g_i)$$

$$= \left(\sum_{k=1}^{n} (h_j, Tg_k)g_k, g_i\right)$$

$$= \sum_{k=1}^{n} (\overline{Tg_k, h_j})(g_k, g_i)$$

$$= (\overline{Tg_i, h_j}).$$

But

$$(Tg_i, h_j) = \left(\sum_{t=1}^{m} a_{ti}h_t, h_j\right)$$

$$= \sum_{t=1}^{m} a_{ti}(h_t, h_j)$$

$$= a_{ji}.$$

We thus have

$$b_{ij} = \overline{a_{ji}},$$

so that (24) is established.

To prove the remaining parts of the theorem we use (23) to conclude from (25) that

$$(Tu,v) = (u,S_1v),$$

for all $u \in U$, $v \in V$. Hence using (23) again, we have

$$(u,T^*v) = (u,S_1v)$$

and thus

$$(u,(T^* - S_1)v) = 0,$$

for all $u \in U$, $v \in V$. If we take $u = (T^* - S_1)v$, it follows that

$$\|(T^* - S_1)v\| = 0$$

and hence

$$(T^* - S_1)v = 0,$$

for all $v \in V$. In other words,

$$S_1 = T^*.$$

As an example, let $U = V_3(R)$, $V = V_2(R)$ and define $T \in L(U,V)$ by

$$T(x_1,x_2,x_3) = (x_1 + 2x_2 + x_3, -x_1 - x_2). \tag{26}$$

We want to write down a formula like (26) for $T^*(x_1,x_2)$. To do this, observe that

$$\begin{aligned}
Te_1 &= T(1,0,0) = (1,-1) = (1,0) - (0,1) \\
Te_2 &= T(0,1,0) = (2,-1) = 2(1,0) - (0,1) \\
Te_3 &= T(0,0,1) = (1,0).
\end{aligned}$$

If we set $g_1 = e_1$, $g_2 = e_2$, $g_3 = e_3$ and $h_1 = (1,0)$, $h_2 = (0,1)$ then G and H are o.n. bases of U and V respectively. Also

$$[T]_G{}^H = \begin{bmatrix} 1 & 2 & 1 \\ -1 & -1 & 0 \end{bmatrix}$$

so that

$$[T^*]_H{}^G = \begin{bmatrix} 1 & -1 \\ 2 & -1 \\ 1 & 0 \end{bmatrix}.$$

Hence

$$\begin{aligned}
T^*(x_1,x_2) &= x_1 T^* h_1 + x_2 T^* h_2 \\
&= x_1(g_1 + 2g_2 + g_3) + x_2(-g_1 - g_2) \\
&= (x_1 - x_2)g_1 + (2x_1 - x_2)g_2 + x_1 g_3 \\
&= (x_1 - x_2, \ 2x_1 - x_2, \ x_1).
\end{aligned}$$

We are now able to define several important classes of linear transformations and matrices in terms of the adjoint and conjugate transpose. Some of the elementary properties of these items will be dealt with now but the main structure theory will be deferred to Chap. 3.

Definition 4.6 (*Classes of transformations and matrices*) *Let U be a finite dimensional unitary space with $T \in L(U,U)$. Let $A \in M_n(R)$, where R is the field of complex numbers. Then*
  (a) *T is **normal** if $TT^* = T^*T$,*
        *A is **normal** if $AA^* = A^*A$,*
  (b) *T is **hermitian** if $T^* = T$,*
        *A is **hermitian** if $A^* = A$.*
  (c) *T is **unitary** if $TT^* = T^*T = I_U$, where I_U is the identity transformation: $I_U u = u$ for all u. The matrix A is **unitary** if $AA^* = A^*A = I_n$.*
  (d) *T is **skew-hermitian** if $T^* = -T$,*
        *A is **skew-hermitian** if $A^* = -A$.*
*If R is the field of real numbers and U is a Euclidean space then these terms become **real normal, symmetric, orthogonal** and **skew-symmetric** respectively.*

Theorem 4.4 *Let U be a finite dimensional unitary or Euclidean space with an o.n. basis $G = \{g_1, \ldots, g_n\}$. Let $T \in L(U,U)$ and suppose $A = [T]_G{}^G$. If U is unitary then T is normal (hermitian, unitary, skew-hermitian) if and only if A has the same property. If U is Euclidean then T is real normal (symmetric, orthogonal, skew-symmetric) if and only if A has the same property.*

Proof. This argument depends directly on the formula (12) and the definition of T^*. For,

$$[T^*]_G{}^G = A^*$$

and hence in case T is normal,

$$AA^* = [T]_G{}^G[T^*]_G{}^G = [TT^*]_G{}^G = [T^*T]_G{}^G = [T^*]_G{}^G[T]_G{}^G = A^*A. \tag{27}$$

Hence A is normal. It is also obvious from the computation that if A is normal, then $[T^*T]_G{}^G = [TT^*]_G{}^G$ and hence $T^*T = TT^*$. We prove one more part of the theorem and leave the rest to the tender mercies of the student (Exercise 6). Suppose then that T is unitary so that $TT^* = T^*T = I_U$. Then since $I_U(g_j) = g_j$, $j = 1, \ldots, n$, we have

$$AA^* = [TT^*]_G{}^G = [I_U]_G{}^G = I_n$$

and

$$A^*A = [T^*T]_G{}^G = [I_U]_G{}^G = I_n.$$

Conversely, if $AA^* = I_n$, then

$$[I_U]_G{}^G = I_n = AA^* = [T]_G{}^G[T^*]_G{}^G = [TT^*]_G{}^G.$$

Hence $I_U = TT^*$ and similarly $T^*T = I_U$.

The various classes of transformations and matrices considered in Def. 4.6 have important geometric characterizations. We shall enunciate them after we prove an interesting result about the inner product (Tu,u). Consider the following example. Let $U = V_2(R)$, R the real-number field, and suppose $T \in L(U,U)$ is defined by

$$Te_1 = e_2, \quad Te_2 = -e_1,$$

where $e_1 = (1,0)$, $e_2 = (0,1)$. Then if $u = a_1e_1 + a_2e_2$ we see, using the standard inner product, that

$$\begin{aligned}
(Tu,u) &= (a_1e_2 - a_2e_1, a_1e_1 + a_2e_2) \\
&= a_1{}^2(e_2,e_1) - a_2{}^2(e_1,e_2) - a_1a_2[(e_1,e_1) - (e_2,e_2)] \\
&= 0.
\end{aligned}$$

Thus (Tu,u) is zero for all $u \in U$ but T is not zero itself. A similar situation cannot happen when R is the complex field.

Definition 4.7   (Quadratic functions)   *Let U and V be finite dimensional unitary (or Euclidean) spaces. If $T \in L(U,V)$ then the **bilinear function** associated with T is the function $\beta(u,v)$ on pairs of vectors $u \in U$, $v \in V$, defined by*

$$\beta(u,v) = (Tu,v). \tag{28}$$

*If $U = V$, then the **quadratic function** $q(u)$ associated with T is the complex valued function on U defined by*

$$q(u) = (Tu,u). \tag{29}$$

Theorem 4.5   *Let $T \in L(U,U)$ where U is a finite dimensional unitary space. Then $T = 0$ if and only if the quadratic function (Tu,u) is zero for every $u \in U$.*

Proof.   As the first step in this proof we observe the simple but important identity:

$$T = \frac{T + T^*}{2} + i\frac{(T - T^*)}{2i}. \tag{30}$$

Let $H = \dfrac{T + T^*}{2}$ and $K = \dfrac{T - T^*}{2i}$. Then clearly $H^* = H$ and $K^* = K$ so that

$$T = H + iK \tag{31}$$

where H and K are hermitian. We see from (31) that

$$(Tu,u) = (Hu,u) + i(Ku,u). \tag{32}$$

Since $H^* = H$, we can conclude from (23) that

$$(\overline{Hu,u}) = (u,Hu) = (Hu,u)$$

and hence (Hu,u) is real. Similarly (Ku,u) is always real. We see then from equation (32) that $(Tu,u) = 0$ if and only if $(Hu,u) = 0$ and

$(Ku,u) = 0$. Thus it suffices to prove the result in the case T is hermitian. Let $G = \{g_1, \ldots, g_n\}$ be an o.n. basis of U and set $A = [T]_G{}^G$. If

$$u = \sum_{i=1}^{n} c_i g_i$$

then

$$(Tu,u) = \left(\sum_{i=1}^{n} c_i T g_i, \sum_{k=1}^{n} c_k g_k \right) \tag{33}$$

$$= \sum_{i,k=1}^{n} c_i \bar{c}_k (T g_i, g_k).$$

According to (15), $(T g_i, g_k) = a_{ki}$. Hence (33) becomes

$$(Tu,u) = \sum_{i,k=1}^{n} c_i \bar{c}_k a_{ki}.$$

Moreover, Theorem 4.4 tells us that $a_{ki} = \bar{a}_{ik}$ (i.e., $A = A^*$). Now define $\varphi(c_1, \ldots, c_n) = (Tu,u)$. The problem is to show that if φ is zero for any choice of $c_1, \ldots, c_n$ then $a_{ki} = 0$, $i,k = 1, \ldots, n$. We compute that

$$\varphi(c_1, \ldots, c_n) = \sum_{i,k=1}^{n} c_i \bar{c}_k a_{ki}$$

$$= \sum_{i=1}^{n} |c_i|^2 a_{ii} + \sum_{i>k=1}^{n} c_i \bar{c}_k a_{ki} + \sum_{1=i<k}^{n} c_i \bar{c}_k a_{ki}$$

$$= \sum_{i=1}^{n} |c_i|^2 a_{ii} + \sum_{i>k=1}^{n} c_i \bar{c}_k a_{ki} + \sum_{i>k=1}^{n} \bar{c}_i c_k a_{ik}$$

$$= \sum_{i=1}^{n} |c_i|^2 a_{ii} + \sum_{i>k=1}^{n} c_i \bar{c}_k a_{ki} + \sum_{i>k=1}^{n} \bar{c}_i c_k \bar{a}_{ki}$$

$$= \sum_{i=1}^{n} |c_i|^2 a_{ii} + \sum_{i>k=1}^{n} (c_i \bar{c}_k a_{ki} + \overline{c_i \bar{c}_k a_{ki}}).$$

First we make the choice $c_i = 0$, $i \neq t$, $c_t = 1$. This yields $a_{tt} = 0$, $t = 1, \ldots, n$. Then for a fixed i_0 and k_0, $i_0 > k_0$, set $c_{i_0} = 1 = c_{k_0}$

and $c_k = 0$, for all k different from i_0 or k_0. Then we conclude that

$$a_{k_0 i_0} + \bar{a}_{k_0 i_0} = 0.$$

(We remark here that with the assumption that U is Euclidean and T symmetric the proof would now be over.) Next set $c_{i_0} = 1$, $c_{k_0} = i$ and $c_k = 0$ for all k not i_0 or k_0. Then

$$-i a_{k_0 i_0} + i \bar{a}_{k_0 i_0} = 0.$$

Thus $a_{k_0 i_0} = 0$. We have proved that $A = 0$ and hence $T = 0$.

By use of Theorem 4.5 we can now establish important geometric properties of those special classes of linear transformations.

Theorem 4.6 *Let U be a finite dimensional unitary space and suppose $T \in L(U,U)$. Then*
  (a)   *T is normal if and only if $\|Tu\| = \|T^*u\|$ for all $u \in U$.*
  (b)   *T is unitary if and only if $\|Tu\| = \|u\|$ for all $u \in U$. This is in turn equivalent to $(Tu, Tv) = (u,v)$ for all u and v in U.*

Proof.   (a)   The equality $\|Tu\| = \|T^*u\|$ implies

$$(Tu, Tu) = (T^*u, T^*u) \text{ or } (T^*Tu, u) = ((T^*)^*T^*u, u).$$

Now $(T^*)^* = T$ follows immediately from the definition of T^* (see Exercise 11) and finally

$$((T^*T - TT^*)u, u) = 0 \tag{34}$$

for all u. Theorem 4.5 then tells us that $T^*T - TT^* = 0$, i.e., that T is normal. We can now reverse the steps starting with (34) to prove that $\|Tu\| = \|T^*u\|$.
  (b)   We first show that $\|Tu\| = \|u\|$ for all $u \in U$ if and only if $(Tu, Tv) = (u,v)$ for all u and v in U. One direction is simple, since if we set $v = u$ we get $(Tu, Tu) = (u,u)$ or $\|Tu\| = \|u\|$. To prove the implication in the other direction let us assume that $\|Tu\| = \|u\|$ for all $u \in U$ and set $u = x + y$. Then

$$(x + y, x + y) = (T(x + y), T(x + y))$$
$$= (Tx, Tx) + (Tx, Ty) + (Ty, Tx) + (Ty, Ty).$$

Thus

$$(x,x) + (x,y) + (y,x) + (y,y)$$
$$= (Tx,Tx) + (Tx,Ty) + (Ty,Tx) + (Ty,Ty).$$

But $(x,x) = (Tx,Tx)$ and $(y,y) = (Ty,Ty)$, and hence we can write

$$(Tx,Ty) - (x,y) = (y,x) - (Ty,Tx)$$

or

$$((T^*T - I_U)x,y) = (y,(I_U - T^*T)x) \tag{35}$$

for any x and y in U. Set $y = (T^*T - I_U)x$ in (35) to obtain

$$\|(T^*T - I_U)x\|^2 = -((I_U - T^*T)x, (I_U - T^*T)x)$$
$$= -\|(I_U - T^*T)x\|^2. \tag{36}$$

But the norm is always nonnegative, and hence $(T^*T - I_U)x = 0$ for all x. Hence

$$((T^*T - I_U)x,y) = 0$$

for all x and $y \in U$; and finally

$$(Tx,Ty) = (x,y)$$

for all x and y in U.

We now prove that the unitary property of T is equivalent to $\|Tu\| = \|u\|$ for all u in U. In one direction, if $T^*T = I_U$ then obviously $\|u\|^2 = (u,u) = (T^*Tu,u) = (Tu,Tu) = \|Tu\|^2$. Conversely, if $(Tu,Tu) = (u,u)$ then $(T^*Tu,u) = (I_Uu,u)$ and $((T^*T - I_U)u,u) = 0$ for all u. We can then apply Theorem 4.5 to conclude that

$$T^*T = I_U. \tag{37}$$

To complete the proof we must show that (37) implies

$$TT^* = I_U. \tag{38}$$

Let $g_1, \ldots, g_n$ be a basis of U. Then $Tg_1, \ldots, Tg_n$ is a basis of

U also, for

$$\sum_{i=1}^{n} c_i T g_i = 0$$

implies that

$$\sum_{i=1}^{n} c_i T^* T g_i = 0.$$

However, (37) tells us that $T^* T g_i = g_i$, and hence

$$\sum_{i=1}^{n} c_i g_i = 0.$$

Thus $T g_1, \ldots, T g_n$ are linearly independent and hence form a basis. Let $h_i = T g_i$, $i = 1, \ldots, n$, and set $H = \{h_1, \ldots, h_n\}$, $G = \{g_1, \ldots, g_n\}$. Then from (37) we have $T^* h_i = T^* T g_i = I_U g_i = g_i$, $i = 1, \ldots, n$. Hence

$$[T^*]_H{}^G = I_n. \tag{39}$$

It is also clear that $[T]_G{}^H = I_n$. Hence from (39) we have

$$[T T^*]_H{}^H = [T]_G{}^H [T^*]_H{}^G = I_n.$$

Thus the statement (38) follows.

As an example, we will illustrate the use of Theorem 4.6 for a particular linear transformation. Thus let $U = V_2(R)$ where R is the complex number field. Let e_1, e_2, be the standard basis and define $T \in L(U, U)$ by

$$Te_1 = \cos \theta \, e_1 + \sin \theta \, e_2$$
$$Te_2 = - \sin \theta \, e_1 + \cos \theta \, e_2$$

where θ is a real number. If $u = c_1 e_1 + c_2 e_2$ is any vector in U then

$$\begin{aligned}
Tu = T(c_1 e_1 + c_2 e_2) &= c_1 T e_1 + c_2 T e_2 \\
&= c_1 (\cos \theta \, e_1 + \sin \theta \, e_2) + c_2 (- \sin \theta \, e_1 + \cos \theta \, e_2) \\
&= (c_1 \cos \theta - c_2 \sin \theta) e_1 + (c_1 \sin \theta + c_2 \cos \theta) e_2.
\end{aligned}$$

Hence

$$\begin{aligned}
\|Tu\|^2 &= |c_1 \cos\theta - c_2 \sin\theta|^2 + |c_1 \sin\theta + c_2 \cos\theta|^2 \\
&= |c_1|^2 \cos^2\theta + |c_2|^2 \sin^2\theta - (c_1\bar{c}_2 + \bar{c}_1 c_2)\cos\theta\sin\theta \\
&\quad + |c_1|^2 \sin^2\theta + |c_2|^2 \cos^2\theta + (c_1\bar{c}_2 + \bar{c}_1 c_2)\cos\theta\sin\theta \\
&= |c_1|^2 (\cos^2\theta + \sin^2\theta) + |c_2|^2 (\cos^2\theta + \sin^2\theta) \\
&= |c_1|^2 + |c_2|^2 \\
&= \|u\|^2.
\end{aligned}$$

Hence $\|Tu\| = \|u\|$ for all $u \in U$, and it follows that T is a unitary transformation.

Quiz

Answer true or false (all vector spaces are finite dimensional over a field R):.

1. The matrix representation of the identity transformation is always the identity matrix.

2. If $U = V_2(R)$, $V = V_4(R)$ and $T \in L(U,V)$ is defined by $T(x_1,x_2) = (x_1,x_2,0,0)$, then $[T]_G^H = I_2$, where G and H are the standard bases in U and V respectively.

3. If $T \in L(U,V)$, $U = V_2(R)$, $V = V_3(R)$, and $T(x_1,x_2) = (x_1 - x_2, 2x_1 - x_2, x_1)$, then $T^*(x_1,x_2,x_3) = (x_1 + 2x_2 + x_3, -x_1 - x_2)$. Use the standard inner products in U and V.

4. The product of two hermitian matrices is hermitian.

5. The quadratic function associated with a hermitian transformation on a unitary space is always real-valued.

6. If R is the field of real numbers, $A \in M_n(R)$, and

$$\varphi(x) = \sum_{i,j=1}^{n} a_{ij} x_i x_j = 0$$

for all vectors $x = (x_1, \ldots, x_n) \in V_n(R)$ then $A = 0$.

7. Suppose U is a unitary space and $q(u) = (Tu,u)$. If $q(g_j) = 0$, $j = 1, \ldots, n$, where $G = \{g_1, \ldots, g_n\}$ is an o.n. basis of U, then $T = 0$.

8. If the product of two hermitian matrices is 0 then one of them must be 0.

9. The set of n-square unitary matrices forms a ring, using the ordinary definitions for matrix addition and multiplication.

10. If U is a unitary space, $T \in L(U,U)$, and $T^*T = 0$ then $T = 0$.

11. The matrix representation of a normal transformation on a unitary space is always a normal matrix.

12. If A is a hermitian matrix and $A^2 = 0$ then $A = 0$.

13. If A and B are n-square hermitian matrices and $A^2 + B^2 = 0$ then $A = B = 0$.

14. If the quadratic function associated with T has a constant value then $T = 0$.
15. If A is hermitian and $(A^*A)_{11} = 0$ then $A_{(1)} = A^{(1)} = 0$.

Exercises

1. Let R be a field and $A \in M_n(R)$. Show that A commutes with every diagonal matrix in $M_n(R)$ if and only if A is a diagonal matrix.
2. Let $T \in L(U,U)$, dim $U = n$. Let W be a subspace of U such that $Tw \in W$ for each $w \in W$, i.e., W is an *invariant subspace* of T. If dim $W = r$, show that there exists a basis $G = \{g_1, \ldots, g_n\}$ of U such that if $A = [T]_G{}^G$ then $a_{ij} = 0$, $i = r + 1, \ldots, n, j = 1, \ldots, r$.
3. Prove formulas (19), (20), and (21).
4. Let T satisfy the conditions of Exercise 2. Show that if U is a unitary space then $W^\perp$ is an invariant subspace of T^*.
5. Show that the product of unitary matrices is unitary.
6. Complete the proof of Theorem 4.4.
7. Let R be the field of complex numbers. If A and B are in $M_n(R)$ show that

$$\sum_{i,j=1}^{n} a_{ij}x_i\bar{x}_j = \sum_{i,j=1}^{n} b_{ij}x_i\bar{x}_j$$

for all $x \in V_n(R)$ implies that $A = B$.
8. Let U be a unitary space. Show that if the quadratic function $q(u) = (Tu,u)$ is real for all $u \in U$ then T is hermitian.
9. Let U be unitary and $T \in L(U,U)$. If $T = H + iK$ where H and K are hermitian show that T is normal if and only if $HK = KH$.
10. Using Exercise 9, show that if $T \in L(U,U)$ is normal and $T^2 = 0$ then $T = 0$.
11. Show that if U and V are unitary and $T \in L(U,V)$ then $(T^*)^* = T$.

1.5   Inverses

In this section we are going to discuss the problem of solving the equation

$$Tu = v \tag{1}$$

for u if v is a given vector and T is a given linear transformation. We make the following definition.

Definition 5.1   (Inverse of a transformation)   *Let U and V be vector spaces over a field R and let $T \in L(U,V)$. If $T(u_1) = T(u_2)$*

implies that $u_1 = u_2$ then T is called **one-one,** *written 1-1. If* rng $T =$ *V then T is said to be* **onto** *V. If T is 1-1 onto V then there exists a function S on V to U defined as follows: $S(v)$ is the unique vector in U such that $T(S(v)) = v$. The function S is called the* **inverse** *of T and is denoted by $S = T^{-1}$. If T is 1-1 and onto V then T is said to be* **nonsingular.** *If T is not nonsingular it is said to be* **singular.**

We assemble in the next theorem a collection of important results about nonsingular transformations.

Theorem 5.1 *Let U and V be vector spaces over a field R and assume that $T \in L(U,V)$. Then the following results hold.*

(a) *If T is nonsingular then $T^{-1} \in L(V,U)$.*

(b) *T is nonsingular if and only if T is onto V and* ker $T = 0$.

(c) *T is nonsingular if and only if*

$$Tu = v$$

has precisely one solution u for each $v \in V$.

(d) *If U is finite dimensional, then T is nonsingular if and only if $\{Tg_1, \ldots, Tg_n\}$ is a basis of V whenever $\{g_1, \ldots, g_n\}$ is a basis of U.*

(e) *If U is finite dimensional, then T is nonsingular if and only if* rng $T = V$ *and there exists an $S \in L(V,U)$ such that*

$$ST = I_U. \tag{2}$$

In fact, the equation (2) implies that $S = T^{-1}$.

Proof. (a) If v_1, v_2 are in V and a_1, a_2 are in R, then there exist u_1 and u_2 in U such that $Tu_1 = v_1$, $Tu_2 = v_2$. By definition, $T^{-1}(v_1) = u_1$ and $T^{-1}(v_2) = u_2$. Moreover, $T(a_1u_1 + a_2u_2) = a_1v_1 + a_2v_2$ and once again $T^{-1}(a_1v_1 + a_2v_2) = a_1u_1 + a_2u_2 = a_1T^{-1}(v_1) + a_2T^{-1}(v_2)$. Thus T^{-1} is linear.

(b) If T is nonsingular then, by definition, T is onto V. Also, if $u \in$ ker T, then $Tu = 0$ and, since $T0 = 0$, it follows from the 1-1 property of T that $0 = u$. Thus ker $T = 0$. Conversely, if ker $T = 0$ then T is 1-1. For, if $Tu_1 = Tu_2$ then $T(u_1 - u_2) = 0$ and hence $u_1 - u_2$ is in ker T. Thus $u_1 = u_2$. If, in addition, T is onto V then T is nonsingular.

(c) By definition, T is nonsingular if and only if each vector $v \in V$ is a value of Tu for precisely one vector $u \in U$.

(d)   Suppose T is nonsingular and $g_1, \ldots, g_n$ is a basis of U. Then $Tg_1, \ldots, Tg_n$ span the range of T and hence must span V. On the other hand if

$$\sum_{i=1}^{n} c_i Tg_i = 0$$

then

$$T\left(\sum_{i=1}^{n} c_i g_i\right) = 0,$$

and T maps

$$\sum_{i=1}^{n} c_i g_i$$

into 0. The nonsingularity of T implies that

$$\sum_{i=1}^{n} c_i g_i = 0$$

and hence $c_1 = \cdots = c_n = 0$. In other words, $Tg_1, \ldots, Tg_n$ is a basis of V. Conversely, suppose that $Tg_1, \ldots, Tg_n$ is a basis of V whenever $g_1, \ldots, g_n$ is a basis of U. Then we can conclude that $\mathrm{rng}\ T = V$. Also if

$$T\left(\sum_{i=1}^{n} c_i g_i\right) = T\left(\sum_{i=1}^{n} d_i g_i\right)$$

then

$$\sum_{i=1}^{n} (c_i - d_i) Tg_i = 0.$$

The linear independence of the Tg_i implies that $c_i = d_i, i = 1, \ldots, n$. We have proved that T is 1-1.

(e) If T is nonsingular then choose $S = T^{-1}$. From the definition of T^{-1} we have

$$T^{-1}Tu = T^{-1}(Tu) = u.$$

Conversely, suppose that rng $T = V$ and there is an $S \in L(V, U)$ such that

$$ST = I_U.$$

Then let $g_1, \ldots, g_n$ be a basis of U. Since rng $T = V$, it follows that $Tg_1, \ldots, Tg_n$ is a spanning set for V. However, if

$$\sum_{i=1}^{n} c_i Tg_i = 0$$

then

$$0 = S \sum_{i=1}^{n} c_i Tg_i = \sum_{i=1}^{n} c_i STg_i = \sum_{i=1}^{n} c_i g_i.$$

The linear independence of the basis vectors $g_1, \ldots, g_n$ implies that $c_1 = \cdots = c_n = 0$. Hence $Tg_1, \ldots, Tg_n$ is a basis of V. By part (d) of this theorem it follows that T is nonsingular. Thus T^{-1} exists satisfying $TT^{-1} = I_V$. It follows from $ST = I_U$ that

$$(ST)T^{-1} = T^{-1}$$

and hence $S = SI_V = S(TT^{-1}) = (ST)T^{-1} = T^{-1}$.

The argument used in the proof of Theorem 5.1 (d) has an important application that simplifies the verification of nonsingularity for linear transformations.

Theorem 5.2 *Assume that U is a finite dimensional vector space over a field R and $T \in L(U, U)$. Then T is nonsingular if and only if there exists an $S \in L(U, U)$ such that*

$$ST = I_U.$$

Moreover, $S = T^{-1}$.

Proof. The point of this theorem is that if the product of two linear transformations on U is the identity on U then both are nonsingular, i.e., have inverses. We shall show by example that T must be in $L(U,U)$ for this to be true.

Let $g_1, \ldots, g_n$ be a basis of U. If

$$\sum_{i=1}^{n} c_i T g_i = 0$$

it follows that

$$S \sum_{i=1}^{n} c_i T g_i = 0,$$

$$\sum_{i=1}^{n} c_i S T g_i = 0,$$

and finally

$$\sum_{i=1}^{n} c_i g_i = 0.$$

Hence $c_1 = \cdots = c_n = 0$, and we conclude that $T g_1, \ldots, T g_n$ are linearly independent. But dim $U = n$ and hence $T g_1, \ldots, T g_n$ constitute a basis of U. In other words, T is onto U and we can apply Theorem 5.1 (e) to finish the proof.

As stated above, it is essential that $T \in L(U,U)$ in order for Theorem 5.2 to hold. For example, let $U = V_2(R)$, $V = V_3(R)$, and set $Te_1 = (1,0,0)$, $Te_2 = (0,1,0)$. If $S \in L(V,U)$ is defined by $S(1,0,0) = e_1$, $S(0,1,0) = e_2$, $S(0,0,1) = e_2$, then

$$ST e_1 = e_1, \quad ST e_2 = e_2$$

and hence $ST = I_U$. But T is not onto V and hence fails to satisfy the definition of nonsingularity.

There is a concept of nonsingularity for matrices that is closely related to the same concept for linear transformations. Before we make the formal definitions, suppose we consider a simple example. Let R be the ring of ordinary integers and set

$$A = \begin{bmatrix} 1 & 1 \\ 1 & 2 \end{bmatrix}.$$

It is very easy to see that if

$$B = \begin{bmatrix} 2 & -1 \\ -1 & 1 \end{bmatrix}$$

then $AB = BA = I_2$. Moreover B has integer entries. Suppose we change A a bit to

$$A_1 = \begin{bmatrix} 1 & 1 \\ -1 & 1 \end{bmatrix}.$$

Let us try to choose B_1 so that $A_1 B_1 = B_1 A_1 = I_2$. Set

$$B_1 = \begin{bmatrix} x & y \\ u & v \end{bmatrix}$$

and write down the conditions that $A_1 B_1 = I_2$. These are

$$x + u = 1$$
$$y + v = 0$$
$$x - u = 0$$
$$v - y = 1.$$

Thus $x = u = \frac{1}{2}$, $y = -v = -\frac{1}{2}$ and

$$B_1 = \begin{bmatrix} \frac{1}{2} & -\frac{1}{2} \\ \frac{1}{2} & \frac{1}{2} \end{bmatrix}.$$

We can directly verify that $A_1 B_1 = B_1 A_1 = I_2$. But notice that the entries of B_1 are no longer integers. In some sense then A_1 does not have an inverse if we insist that the entries of the inverse be integers.

Definition 5.2 (*Unit matrices*) *Let R be a commutative ring with an identity for multiplication, denoted by 1. An n-square matrix A with entries in R is called a **unit** in $M_n(R)$ if there exists a matrix $B \in M_n(R)$ such that*

$$AB = BA = I_n. \tag{3}$$

*A is also said to be **nonsingular** if it is a unit. If A is not a unit in $M_n(R)$, it is said to be **singular**.*

According to this definition, if R is the ring of integers then the matrix A in the above example is a unit whereas the matrix A_1 is not.

The first thing we remark is that if $AB = BA = I_n$, as in (3), and there exists a matrix $B_1 \in M_n(R)$ such that $AB_1 = B_1A = I_n$ then $B_1 = B_1 I_n = B_1(AB) = (B_1 A)B = I_n B = B$. In other words, there is at most one matrix B satisfying (3). It is also clear that if $A_1, \ldots, A_p$ are unit matrices in $M_n(R)$ then the product $A_1 \cdots A_p$ is a unit matrix also and

$$(A_1 A_2 \cdots A_p)^{-1} = A_p^{-1} A_{p-1}^{-1} \cdots A_1^{-1}.$$

For,

$$(A_p^{-1} A_{p-1}^{-1} \cdots A_1^{-1})(A_1 A_2 \cdots A_p)$$
$$= (A_p^{-1} \cdots A_2^{-1})(A_1^{-1} A_1)(A_2 \cdots A_p)$$
$$= (A_p^{-1} \cdots A_2^{-1})(A_2 \cdots A_p)$$
$$= \cdots$$
$$= I_n.$$

Definition 5.3 (*Inverse of a unit matrix*)   *Let R be a commutative ring with a multiplicative identity 1. If A is a unit in $M_n(R)$ then the unique matrix $B \in M_n(R)$ for which (3) holds is called the **inverse** of A. It is denoted by $B = A^{-1}$.*

There is a very interesting connection between the nonsingularity of a linear transformation and the nonsingularity of any matrix representation of it.

Theorem 5.3   *Let U be a finite dimensional vector space over a field R, dim $U = n$. Suppose that $G = \{g_1, \ldots, g_n\}$ and $H = \{h_1, \ldots, h_n\}$ are bases of U and $T \in L(U,U)$. Set $A = [T]_H{}^G \in M_n(R)$. Then T is nonsingular if and only if A is a unit in $M_n(R)$. Moreover,*

$$[T^{-1}]_G{}^H = A^{-1}. \tag{4}$$

Proof.   Suppose first that T is nonsingular so that

$$TT^{-1} = T^{-1}T = I_U.$$

Then, by Theorem 4.1, we know that

$$I_n = [I_U]_H{}^H = [T^{-1}T]_H{}^H = [T^{-1}]_G{}^H[T]_H{}^G$$

and similarly

$$I_n = [I_U]_G{}^G = [TT^{-1}]_G{}^G = [T]_H{}^G[T^{-1}]_G{}^H.$$

If we set $B = [T^{-1}]_G{}^H$ then we have proved that

$$I_n = BA = AB.$$

Hence

$$B = A^{-1}.$$

Conversely, if A is a unit and we define $S \in L(U,U)$ by $[S]_G{}^H = A^{-1}$ then, once again by Theorem 4.1,

$$\begin{aligned}
[ST]_H{}^H &= [S]_G{}^H[T]_H{}^G \\
&= A^{-1}A = I_n.
\end{aligned}$$

Hence $ST = I_U$ and we conclude the result by applying Theorem 5.2.

We are now in a position to answer the following important question. Suppose that G and H are two bases for U, dim $U = n$, and $T \in L(U,U)$. What is the relation between the two n-square matrices that are matrix representations of T with respect to G and H? The answer is given in

Theorem 5.4 *Let $G = \{g_1, \ldots, g_n\}$ be a basis of the vector space U over the field R. Suppose that $T \in L(U,U)$ and $A = [T]_G{}^G$. If B is any matrix in $M_n(R)$ then there exists a basis $H = \{h_1, \ldots, h_n\}$ of U such that $[T]_H{}^H = B$ if and only if there exists a unit matrix $C \in M_n(R)$ such that*

$$A = CBC^{-1}. \tag{5}$$

Proof. Suppose a basis H exists such that $B = [T]_H{}^H$. Then, by

Theorem 4.1, we compute that

$$A = [T]_G^G = [I_U T I_U]_G^G = [I_U]_H^G [T]_H^H [I_U]_G^H. \qquad (6)$$

Now set $C = [I_U]_H^G$ and observe that $I_n = [I_U]_G^G = [I_U]_H^G [I_U]_G^H$ and $I_n = [I_U]_H^H = [I_U]_G^H [I_U]_H^G$.

Hence $C^{-1} = [I_U]_G^H$. We conclude from (6) that

$$A = CBC^{-1}.$$

To prove the converse, suppose that a unit matrix $C \in M_n(R)$ exists such that $A = CBC^{-1}$. Define vectors $h_1, \ldots, h_n$ by

$$h_j = \sum_{i=1}^{n} c_{ij} g_i, \quad j = 1, \ldots, n. \qquad (7)$$

We will prove that $h_1, \ldots, h_n$ are linearly independent and therefore form a basis. Suppose now that

$$\sum_{j=1}^{n} d_j h_j = 0, \quad d_j \in R, \quad j = 1, \ldots, n.$$

Then from (7) we have

$$0 = \sum_{j=1}^{n} d_j h_j = \sum_{j=1}^{n} d_j \sum_{i=1}^{n} c_{ij} g_i$$

$$= \sum_{i=1}^{n} \left(\sum_{j=1}^{n} c_{ij} d_j \right) g_i.$$

The vectors $g_1, \ldots, g_n$ are linearly independent and hence

$$\sum_{j=1}^{n} c_{ij} d_j = 0, \quad i = 1, \ldots, n. \qquad (8)$$

If we set

$$d = \begin{bmatrix} d_1 \\ \cdot \\ \cdot \\ \cdot \\ d_n \end{bmatrix} \in M_{n,1}(R)$$

then the equation (8) can be simply stated as

$$Cd = 0. \tag{9}$$

But C^{-1} exists, and hence

$$C^{-1}(Cd) = 0,$$

or

$$d = 0.$$

We have proved that $d_1 = \cdots = d_n = 0$ and thus $h_1, \ldots, h_n$ are linearly independent. It follows that $H = \{h_1, \ldots, h_n\}$ is a basis of U. We want to compute $[T]_H{}^H$. Using Theorem 4.1 again, we compute that

$$[T]_H{}^H = [I_U]_G{}^H[T]_G{}^G[I_U]_H{}^G. \tag{10}$$

But

$$I_U h_j = h_j = \sum_{i=1}^{n} c_{ij} g_i$$

and hence

$$[I_U]_H{}^G = C. \tag{11}$$

As we saw earlier, (11) implies that

$$C^{-1} = [I_U]_G{}^H. \tag{12}$$

Thus, by looking at (5), (10) becomes

$$[T]_H{}^H = C^{-1}AC = B.$$

The problem of systematically computing the inverse of a matrix will be deferred to Chap. 2. In certain special cases, however, we can write down the inverse of a matrix with ease. For example, if R is the complex field and A is an n-square unitary matrix it follows by definition that

$$AA^* = A^*A = I_n.$$

Hence for A unitary

$$A^{-1} = A^*. \tag{13}$$

Definition 5.4   **(Similarity)**   *Let R be a field and A and B be matrices in $M_n(R)$. Then A is **similar** to B **over** R if there exists a unit matrix $C \in M_n(R)$ such that*

$$A = CBC^{-1}. \tag{14}$$

We can restate Theorem 5.4 as follows.

Theorem 5.5   *Let U be a finite dimensional vector space over the field R. Two matrices in $M_n(R)$ are matrix representations of the same linear transformation in $L(U,U)$ if and only if they are similar over R.*

Quiz

Answer true or false (all vector spaces are finite dimensional over a field R):

1. The matrix $\begin{bmatrix} 1 & 1 \\ 1 & 1 \end{bmatrix}$ is singular in $M_2(R)$ where R is the ring of ordinary integers.
2. If $T \in L(U,V)$ and $\eta(T) \geq 1$, then T is singular.
3. If T and S are in $L(U,U)$ and TS is singular, then both T and S are singular.
4. If R is the real field, then $\begin{bmatrix} 1 & 2 \\ 3 & 4 \end{bmatrix}$ and $\begin{bmatrix} 4 & 2 \\ 3 & 1 \end{bmatrix}$ are similar matrices.
5. If R is the field of complex numbers and $A \in M_n(R)$ is unitary and hermitian, then $A^2 = I_n$.
6. If $A = \text{diag}(\alpha_1, \ldots, \alpha_n)$, then A is singular if and only if

$$\prod_{i=1}^{n} \alpha_i = 0.$$

7. If G and H are bases for U, then $[I_U]_G{}^H$ is the inverse of $[I_U]_G{}^H$.
8. If $A \in M_n(R)$ and A has a zero row, then A is singular.
9. If U is unitary and $T \in L(U,U)$ is nonsingular, then $\|Tu\| = 0$ implies $u = 0$.
10. If A and B are in $M_n(R)$ and $AB + BA = 0$, then at least one of A and B is singular.

Exercises

All vector spaces are finite dimensional over a field R.

1. Show that if $T \in L(U,V)$ and $\rho(T) < \dim V$, then T cannot be nonsingular.

2. If $U = V = V_3(R)$ and $T \in L(U,V)$ is defined by

$$T(x_1,x_2,x_3) = (x_1 + x_2,\ x_2 + x_3,\ x_3 + x_1)$$

then find a similar formula for T^{-1}.

3. Show that if $T \in L(U,U)$ is nonsingular and $g_1, \ldots, g_r$ are linearly independent vectors then $Tg_1, \ldots, Tg_r$ are linearly independent as well.

4. Show that if $T \in L(U,U)$ and T is singular then there exists a nonzero vector $u \in U$ such that $Tu = 0$.

5. Show that if R is the ring of ordinary integers and $A \in M_2(R)$ is given by

$$A = \begin{bmatrix} a_1 & a_2 \\ a_3 & a_4 \end{bmatrix}$$

then A is a unit in $M_2(R)$ if and only if $a_1a_4 - a_2a_3 = \pm 1$.

6. Let $C \in M_r(R)$ be a unit, R a field. Let $g_1, \ldots, g_r$ be linearly independent in U and define

$$h_j = \sum_{i=1}^{r} c_{ij}g_i, \quad j = 1, \ldots, r.$$

Then prove that $h_1, \ldots, h_r$ are linearly independent.

7. Let $g_1, \ldots, g_n$ and $h_1, \ldots, h_n$ be bases of U. Let $C \in M_n(R)$ and suppose that

$$h_j = \sum_{i=1}^{n} c_{ij}g_i, \quad j = 1, \ldots, n.$$

Show that C is a unit in $M_n(R)$.

8. Let $T \in L(U,U)$. Show that if T is nonsingular and $G = \{g_1, \ldots, g_n\}$ is any basis of U, then there exists a basis H of U such that $[T]_G^H = I_n$.

9. For the linear transformation of Exercise 2, find the solution u to the equation $Tu = (1,1,1)$.

10. State whether the following two matrices can represent the same linear transformation:

$$A = (a_{ij}), \quad \begin{cases} a_{ij} = 1 \text{ for } j = i + 1,\ i = 1,\ \ldots,\ n - 1 \\ a_{ij} = 0 \text{ otherwise.} \end{cases}$$

$$B = (b_{ij}), \quad \begin{cases} b_{ij} = 1 \text{ for } j = i + 1,\ i = 1,\ \ldots,\ n - 1, \\ b_{n1} = 1, \\ b_{ij} = 0 \text{ otherwise.} \end{cases}$$

Why?

11. Let $d \in V_n(R)$ and $C \in M_n(R)$. Show that if C is nonsingular and $Cd = 0$, then $d = 0$.

CHAPTER 2

Linear Equations
and Determinants

2.1 Rank of a Matrix

In this section we will begin to study the problem of finding all vectors $u \in U$ for which

$$Tu = v. \tag{1}$$

Here $T \in L(U,V)$ and v is a fixed vector in V. The vector spaces U and V are finite dimensional over the same field R. To compute such vectors u we will introduce bases for U and V. Thus, let $G = \{g_1, \ldots, g_n\}$ and $H = \{h_1, \ldots, h_m\}$ be bases of U and V respectively. Set $A = [T]_G{}^H \in M_{m,n}(R)$. If

$$u = \sum_{i=1}^{n} x_i g_i \quad \text{and} \quad v = \sum_{s=1}^{m} b_s h_s,$$

then

$$Tu = T \sum_{i=1}^{n} x_i g_i = \sum_{i=1}^{n} x_i T g_i \tag{2}$$

$$= \sum_{i=1}^{n} x_i \sum_{s=1}^{m} a_{si} h_s$$

$$= \sum_{s=1}^{m} \left(\sum_{i=1}^{n} a_{si} x_i \right) h_s.$$

A necessary and sufficient condition that $Tu = v$ then reduces to

$$\sum_{i=1}^{n} a_{si} x_i = b_s, \quad s = 1, \ldots, m. \tag{3}$$

If we let x denote the $n \times 1$ matrix $[x_1, \ldots, x_n]^T$ and b denote the $m \times 1$ matrix $[b_1, \ldots, b_m]^T$, then (3) takes the particularly simple form

$$Ax = b. \tag{4}$$

There is a standard language that is used about these items: the n-tuple or $n \times 1$ matrix x is called the *coordinate vector* for u. It, of course, depends on the basis G. Although there is a logical difference between $V_n(R)$ and $M_{n,1}(R)$ we shall not distinguish between the two in studying the equation (4). The statement (4) is called a *system of linear equations*. The system is said to be *homogeneous* if $b = 0$. The set of solutions $x = (x_1, \ldots, x_n) \in V_n(R)$ to the homogeneous system $Ax = 0$ is called the *null space* of A. It is an easy exercise (Exercise 10) to see that in fact the null space of A is a subspace of $V_n(R)$. (See proof of Theorem 2.2, Chap. 1.)

We can summarize our little calculations so far in

Theorem 1.1 Let $T \in L(U,V)$ and let $G = \{g_1, \ldots, g_n\}$, $H = \{h_1, \ldots, h_m\}$ be bases for U and V respectively. Set $A = [T]_G^H \in M_{m,n}(R)$. If x and b are the coordinate vectors for u and v respectively, then

$$Tu = v$$

if and only if

$$Ax = b.$$

This theorem will be used in actually solving the problem indicated in (1). In Chap. 3 we will study a rather more complicated question. Let $T \in L(U,U)$. For what scalars $r \in R$ and nonzero vectors u does

$$Tu = ru \qquad (5)$$

have a solution? It might seem that this question is simpler than solving (1) because v is a multiple of u. What makes it harder in fact is finding the scalar r.

In Sec. 1.2 we defined the rank of a linear transformation and we proved that the existence of a solution to a system of linear equations depends on the rank of a transformation defined in terms of the coefficients of the system. We are now going to give an independent definition of the rank of a matrix and eventually prove that the two concepts are equivalent, i.e., that the rank of a linear transformation $T \in L(U,V)$ is equal to the rank of $A = [T]_G{}^H$ for any choice of bases G and H of U and V respectively.

Definition 1.1 (Row rank, column rank) *Let $A \in M_{m,n}(R)$. The space $R(A) = \langle A_{(1)}, \ldots , A_{(m)} \rangle$ is called the **row space** of A and the space $C(A) = \langle A^{(1)}, \ldots , A^{(n)} \rangle$ is called the **column space** of A. The dimensions of $R(A)$ and $C(A)$ are called the **row rank** of A and the **column rank** of A, respectively. We can obviously think of $A_{(i)}$ as either being in $V_n(R)$ or in $M_{1,n}(R)$ and thus $R(A)$ can be regarded as a subspace of $V_n(R)$ or $M_{1,n}(R)$. Similarly, $C(A)$ can be regarded either as a subspace of $V_m(R)$ or $M_{m,1}(R)$.*

Theorem 1.2 *The row rank of a matrix is equal to its column rank.*

Proof. Let $A = (a_{ij}) \in M_{m,n}(R)$ and denote the row rank and the column rank of any matrix X by $r(X)$ and $c(X)$ respectively. We must prove that

$$r(A) = c(A).$$

For brevity, denote $r(A)$ by r. Let the rows of A numbered $\omega_1, \ldots , \omega_r$ form a basis for the row space of A. Consider the matrix $B \in M_{r,n}(R)$ composed of these rows, i.e.,

$$B_{(i)} = A_{(\omega_i)}, \quad i = 1, \ldots , r. \qquad (6)$$

Then

$$r(B) = \dim \langle B_{(1)}, \ldots, B_{(r)} \rangle$$
$$= \dim \langle A_{(\omega_1)}, \ldots, A_{(\omega_r)} \rangle$$
$$= \dim \langle A_{(1)}, \ldots, A_{(m)} \rangle$$
$$= r.$$

We assert that $c(B) \geq c(A)$. The rows $A_{(\omega_1)}, \ldots, A_{(\omega_r)}$ form a basis for the row space, and therefore

$$A_{(i)} = \sum_{t=1}^{r} \alpha_{it} A_{(\omega_t)} \tag{7}$$

where $\alpha_{it} \in R$, $t = 1, \ldots, r$, $i = 1, \ldots, m$. Elementwise, (7) becomes

$$a_{ij} = \sum_{t=1}^{r} \alpha_{it} a_{\omega_t j}, \quad i = 1, \ldots, m, \quad j = 1, \ldots, n.$$

Now suppose that

$$\sum_{j=1}^{n} \beta_j B^{(j)} = 0, \quad \beta_j \in R, \quad j = 1, \ldots, n, \tag{8}$$

or

$$\sum_{j=1}^{n} \beta_j a_{\omega_i j} = 0, \quad i = 1, \ldots, r. \tag{9}$$

For $k = 1, \ldots, m$,

$$\sum_{j=1}^{n} \beta_j a_{kj} = \sum_{j=1}^{n} \beta_j \sum_{t=1}^{r} \alpha_{kt} a_{\omega_t j}$$
$$= \sum_{t=1}^{r} \alpha_{kt} \left(\sum_{j=1}^{n} \beta_j a_{\omega_t j} \right) = 0,$$

by (9). Thus

$$\sum_{j=1}^{n} \beta_j A^{(j)} = 0. \tag{10}$$

Hence any linear dependence relation (8) among the columns of B implies a similar linear dependence relation (10) among the columns of A. It follows that

$$c(B) \geq c(A).$$

To see this latter inequality, suppose $A^{(i_1)}, \ldots, A^{(i_p)}$ constitute a basis of $C(A)$. If $B^{(i_1)}, \ldots, B^{(i_p)}$ were linearly dependent then we could obtain scalars $\beta_{i_1}, \ldots, \beta_{i_p}$ not all 0, such that

$$\sum_{t=1}^{p} \beta_{i_t} B^{(i_t)} = 0.$$

Let $\beta_j = 0, j \neq i_1, \ldots, i_p$, so that

$$\sum_{j=1}^{n} \beta_j B^{(j)} = 0.$$

It follows that

$$\sum_{j=1}^{n} \beta_j A^{(j)} = 0$$

and hence that

$$\sum_{t=1}^{p} \beta_{i_t} A^{(i_t)} = 0.$$

This is impossible because not all $\beta_{i_1}, \ldots, \beta_{i_p}$ are zero and $A^{(i_1)}, \ldots, A^{(i_p)}$ are linearly independent. It follows that $B^{(i_1)}, \ldots, B^{(i_p)}$ are linearly independent, and thus $c(B) \geq p = c(A)$.

Now $B \in M_{r,n}(R)$, and thus $\langle B^{(1)}, \ldots, B^{(n)} \rangle \subset V_r(R)$. Thus, by Theorem 1.3, Chap. 1,

$$c(B) = \dim \langle B^{(1)}, \ldots, B^{(n)} \rangle \leq \dim V_r(R) = r.$$

Therefore,

$$r(A) = r \geq c(B) \geq c(A). \tag{11}$$

In words, the column rank of any matrix A does not exceed its row rank. Apply this result to A^T,

$$r(A^T) \geq c(A^T).$$

But clearly

$$r(A^T) = c(A) \text{ and } c(A^T) = r(A).$$

Therefore

$$c(A) \geq r(A) \tag{12}$$

and, from (11) and (12),

$$c(A) = r(A).$$

Definition 1.2 (*Rank of a matrix*) *The common value of the row rank and the column rank of a matrix A is called the **rank** of A and is denoted by $\rho(A)$.*

An immediate consequence of this definition is that

$$\rho(A) = \rho(A^T). \tag{13}$$

Before we can prove three important results on the rank of a matrix we need the following useful theorem on the product of two matrices.

Theorem 1.3 *If $A \in M_{m,n}(R)$ and $B \in M_{n,p}(R)$ then*

$$(AB)_{(i)} = A_{(i)}B = \sum_{t=1}^{n} a_{it}B_{(t)}$$

and

$$(AB)^{(j)} = AB^{(j)} = \sum_{t=1}^{n} b_{tj}A^{(t)}.$$

Proof. The matrices $(AB)_{(i)}$, $A_{(i)}B$, and $\sum_{t=1}^{n} a_{it}B_{(t)}$ are $1 \times p$. We show that the $(1, j)$ entry in each of them is equal to $\sum\limits_{t=1}^{n} a_{it}b_{tj}$:

$$[(AB)_{(i)}]_{1,j} = (AB)_{i,j} = \sum_{t=1}^{n} a_{it}b_{tj};$$

$$(A_{(i)}B)_{1,j} = \sum_{t=1}^{n} (A_{(i)})_{1,t}b_{tj} = \sum_{t=1}^{n} a_{it}b_{tj};$$

$$\Big(\sum_{t=1}^{n} a_{it}B_{(t)}\Big)_{1,j} = \sum_{t=1}^{n} a_{it}(B_{(t)})_{1,j} = \sum_{t=1}^{n} a_{it}b_{tj}.$$

Now, using formula (20) of Sec. 1.4, we have

$$\begin{aligned}
[(AB)^{(j)}]^T &= [(AB)^T]_{(j)} \\
&= (B^TA^T)_{(j)} \\
&= B^T_{(j)}A^T \\
&= (B^{(j)})^TA^T \\
&= (AB^{(j)})^T.
\end{aligned}$$

Therefore

$$(AB)^{(j)} = AB^{(j)}.$$

Similarly,

$$\begin{aligned}
[(AB)^{(j)}]^T &= (B^TA^T)_{(j)} \\
&= \sum_{t=1}^{n} (B^T)_{jt}(A^T)_{(t)} \\
&= \sum_{t=1}^{n} b_{tj}(A^{(t)})^T,
\end{aligned}$$

and therefore

$$(AB)^{(j)} = \sum_{t=1}^{n} b_{tj}A^{(t)}.$$

Theorem 1.4 *The rank of the product of two matrices cannot exceed the rank of either factor.*

Proof. Let $A \in M_{m,n}(R)$ and $B \in M_{n,p}(R)$. Then, by Theorem 1.3,

$$\langle (AB)_{(1)}, \ldots, (AB)_{(m)} \rangle = \Big\langle \sum_{t=1}^{n} a_{1t}B_{(t)}, \ldots, \sum_{t=1}^{n} a_{mt}B_{(t)} \Big\rangle$$
$$\subset \langle B_{(1)}, \ldots, B_{(n)} \rangle$$

and thus, by Theorem 1.3, Chap. 1,

$$\rho(AB) = \dim \langle (AB)_{(1)}, \ldots, (AB)_{(m)} \rangle$$
$$\leq \dim \langle B_{(1)}, \ldots, B_{(n)} \rangle = \rho(B).$$

Hence the rank of a product of two matrices cannot exceed the rank of the second factor. But, by (13),

$$\rho(AB) = \rho[(AB)^T] = \rho(B^T A^T) \leq \rho(A^T) = \rho(A).$$

Theorem 1.5   *If $P \in M_m(R)$ and $Q \in M_n(R)$ are nonsingular matrices and A is any matrix in $M_{m,n}(R)$, then*

$$\rho(A) = \rho(PA) = \rho(AQ) = \rho(PAQ).$$

Proof.   By Theorem 1.4,

$$\rho(PA) \leq \rho(A).$$

Also, by the same theorem,

$$\rho(A) = \rho[P^{-1}(PA)] \leq \rho(PA).$$

Therefore

$$\rho(A) = \rho(PA).$$

Now, Q^T is nonsingular (i.e., $QQ^{-1} = I_n$ implies $I_n = (QQ^{-1})^T = (Q^{-1})^T Q^T$, etc.) and hence

$$\rho(AQ) = \rho[(AQ)^T] = \rho(Q^T A^T) = \rho(A^T) = \rho(A).$$

Finally,

$$\rho(PAQ) = \rho[P(AQ)] = \rho(AQ) = \rho(A).$$

Theorem 1.6   *If $T \in L(U,V)$ and $A = [T]_G{}^H$ for any given bases G of U and H of V, then*

$$\rho(T) = \rho(A).$$

Proof. Suppose that the nullity of T is greater than 0. Let u_{r+1}, ..., u_n form a basis for ker T and complete it to a basis $G' = \{u_1, \ldots, u_n\}$ of U. We showed in the proof of Theorem 2.1, Chap. 1, that $v_1 = T(u_1), \ldots, v_r = T(u_r)$ form a basis of rng T. Complete these vectors to a basis $H' = \{v_1, \ldots, v_m\}$ of V. Then

$$T(u_j) = \begin{cases} v_j, & j = 1, \ldots, r, \\ 0, & j = r+1, \ldots, n. \end{cases}$$

Thus $[T]_{G'}{}^{H'} = B \in M_{m,n}(R)$, where

$$B_{ij} = \begin{cases} \delta_{ij}, & i = 1, \ldots, r, \\ 0, & \text{otherwise} \end{cases}$$

i.e.,

$$B_{(i)} = \begin{cases} e_i, & i = 1, \ldots, r, \\ 0, & i = r+1, \ldots, m, \end{cases}$$

where $e_1, e_2, \ldots$ are the standard basis vectors in Def. 1.3, Chap. 1. Clearly,

$$\rho(B) = r = \rho(T).$$

If the nullity of T is 0, just take any basis $G' = \{u_1, \ldots, u_n\}$ of U and complete $v_1 = T(u_1), \ldots, v_n = T(u_n)$ to a basis $H' = \{v_1, \ldots, v_m\}$ of V. Then by an argument identical to the above, we get

$$B_{ij} = \begin{cases} \delta_{ij}, & i = 1, \ldots, n, \\ 0, & \text{otherwise} \end{cases}$$

and

$$\rho(B) = n = \rho(T).$$

Now let G and H be any bases of U and V respectively and let $A = [T]_G{}^H$. Then setting $T = I_V T I_U$ we have

$$[T]_G{}^H = [I_V]_{H'}{}^H [T]_{G'}{}^{H'} [I_U]_G{}^{G'}$$

where $[I_U]_G{}^{G'}$ and $[I_V]_{H'}{}^H$ are matrices of identity transformations and are therefore nonsingular. In fact, as we saw in the proof of Theorem 5.4, Chap. 1, the inverse of $[I_U]_{G'}{}^G$ is $[I_U]_G{}^{G'}$ and similarly the inverse of $[I_V]_{H'}{}^H$ is $[I_V]_H{}^{H'}$. Thus, by Theorem 1.5,

$$\rho([T]_G{}^H) = \rho([T]_{G'}{}^{H'})$$

or

$$\rho(A) = \rho(B).$$

Therefore

$$\rho(A) = \rho(T).$$

Quiz

Answer true or false:

1. If A, B, and C are in $M_n(R)$, $C \neq 0$, and $\rho(A) > \rho(B)$ then $\rho(AC) \geq \rho(BC)$.
2. For any $A \in M_n(R)$, $\rho(A) \leq \rho(A^2) + 1$.
3. $\rho(A + B) \leq \rho(A) + \rho(B)$.
4. If $\rho(A) < \rho(B)$ then $\rho(A) \leq \rho(AB) \leq \rho(B)$.
5. If $T \in L(U,V)$ and ker $T = 0$ then $\rho([T]_G{}^H) = \dim U$ for any bases G and H.
6. $\rho([I_U]_G{}^H) = \dim U$ for any bases H and G of U.
7. If $T \in L(U,V)$ and $Tu = 0$ for all $u \in U$, then $\rho([T]_G{}^H) = 0$.
8. $\rho(AB) \leq \rho(A)\rho(B)$.
9. Let $\{g_1, \ldots, g_n\}$ be a basis for U, and let $T \in L(U,U)$ be defined by $Tg_1 = g_2$, $Tg_2 = g_3$, $\ldots$, $Tg_{n-1} = g_n$, $Tg_n = 0$. Then $\rho[([T]_H{}^H)^n] = 0$ for any basis H of U.
10. If $\rho(AB) = \rho(A) = \rho(B) > 0$, where A and B are n-square matrices, then either A or B is nonsingular.

Exercises

1. Let $G = \{u_1, \ldots, u_n\}$ and $H = \{v_1, \ldots, v_m\}$ be bases for U and V respectively. Let $T_{ji} \in L(U,V)$ be defined as follows:

$$T_{ji}(u_s) = \delta_{js}v_i, \quad s = 1, \ldots, n.$$

[See proof of Theorem 2.1, Chap. 1, formula (10), etc.] Show that $[T_{ji}]_G{}^H$ is the $m \times n$ matrix with 1 in its (i,j) position and 0 elsewhere. The standard notation for such a matrix is E_{ij}.

2. Let $E_{ij} \in M_n(R)$, $i, j = 1, \ldots, n$, be the matrices defined in Exercise 1. Show that the matrices E_{ij}, $i, j = 1, \ldots, n$, form a basis for the space $M_n(R)$. Prove the multiplication formula

$$E_{pq}E_{rs} = \delta_{qr}E_{ps}$$

and deduce the multiplication formula for any A and B in $M_n(R)$.

3. Prove that

$$\rho(E_{ij} - E_{hk}) = 2 - \delta_{ih} - \delta_{jk}.$$

4. Let $T \in L(V,V)$, dim $V = n$, and suppose that $\rho(T^k) = \rho(T^{k+1})$ for some positive integer k. Prove that $\rho(T^i) = \rho(T^j)$ for any integers $i \geq k$ and $j \geq k$.
5. Prove that $\rho(A^n) = \rho(A^{n+1})$ for any $A \in M_n(R)$.
6. If $A \in M_n(R)$ and $\rho(A) < n$ show that there exists a nonzero matrix $B \in M_n(R)$ such that $AB = 0$.
7. Let $T: V_4(R) \to V_6(R)$ be a linear transformation defined by $T(x_1,x_2,x_3,x_4) = (x_1 - x_2, x_1 - x_3, x_1 - x_4, x_2 - x_3, x_2 - x_4, x_3 - x_4)$ for all $(x_1,x_2,x_3,x_4) \in V_4(R)$. Construct $[T]_G^H$ where G and H are the standard bases in $V_4(R)$ and $V_6(R)$ respectively. Hence, or otherwise, find the rank of T.
8. Show that $A \in M_{m,n}(R)$ has rank 1 if and only if there exist nonzero $x \in M_{m,1}(R)$ and $y \in M_{1,n}(R)$ such that $A = xy$.
9. Let $E_{ij} \in M_{m,n}(R)$ be the matrix defined in Exercise 1. Show that $\rho(E_{ij}) = 1$ and find $x \in M_{m,1}(R)$ and $y \in M_{1,n}(R)$ such that $E_{ij} = xy$.
10. Show that the null space of a matrix $A \in M_{m,n}(R)$ is a subspace of $V_n(R)$.

2.2 Operations on Matrices

We shall now discuss techniques which will produce a method for completely solving systems of linear equations. These techniques will also yield a constructive procedure for finding the rank of a given matrix.

Two systems of linear equations

$$\sum_{j=1}^{n} a_{ij}x_j = b_i, \quad i = 1, \ldots, m, \tag{1}$$

and

$$\sum_{j=1}^{n} a'_{ij}x_j = b'_i, \quad i = 1, \ldots, k, \tag{2}$$

are said to be *equivalent* if any solution of (1) is a solution of (2) and vice versa. In other words, (1) and (2) are equivalent if, for any n-tuple $c = (c_1, \ldots, c_n)$ satisfying

$$\sum_{j=1}^{n} a_{ij}c_j = b_i, \quad i = 1, \ldots, m,$$

we have

$$\sum_{j=1}^{n} a'_{ij}c_j = b'_i, \quad i = 1, \ldots, k,$$

and vice versa.

The systems (1) and (2) can also be written in matrix notation

$$Ax = b \tag{3}$$

and

$$A'x = b'. \tag{4}$$

They are equivalent if $Ac = b$ implies $A'c = b'$, and vice versa.

The methods for solving a system of linear equations (1) depend on successively replacing (1) by simpler equivalent systems. For example, to solve the system

$$x_1 + 6x_2 + 3x_3 + 4x_4 = 1$$
$$x_1 + 2x_2 + x_3 + x_4 = 1$$
$$-x_1 + 2x_2 + x_3 + 2x_4 = -1$$

we can simplify by adding each side of the last equation to the corresponding sides of each of the preceding equations. We obtain the obviously equivalent system

$$8x_2 + 4x_3 + 6x_4 = 0$$
$$4x_2 + 2x_3 + 3x_4 = 0$$
$$-x_1 + 2x_2 + x_3 + 2x_4 = -1.$$

We now add -2 times the second equation to the first producing

$$0x_2 + 0x_3 + 0x_4 = 0$$
$$4x_2 + 2x_3 + 3x_4 = 0$$
$$-x_1 + 2x_2 + x_3 + 2x_4 = -1.$$

We can, of course, ignore the first of these. We multiply the second equation by $\frac{1}{4}$ and the third one by -1 and add 2 times the second to the third:

$$x_2 + \tfrac{1}{2}x_3 + \tfrac{3}{4}x_4 = 0$$
$$x_1 - \tfrac{1}{2}x_4 = 1.$$

This is essentially the simplest equivalent system: we can take any values for x_3 and x_4 and compute the corresponding values for x_1 and x_2.

The following operations obviously change a system of linear equations into an equivalent system:

(I)   interchange two equations;
(II) add to each side of an equation the corresponding sides of another equation multiplied by a scalar;
(III) multiply both sides of an equation by a nonzero scalar.

If the system is written in the form (3) these operations on the left-hand sides of the equations correspond to operations on rows of the matrix A. In other words, the unknowns $x_1, \ldots, x_n$ do not have to be dragged along through the reduction process.

Definition 2.1 (*Elementary row operations*) *The three elementary row operations on a matrix $A \in M_{m,n}(R)$ are:*

(I)   *interchange the i-th and the j-th row of A;*
(II)   *add c times the j-th row of A to the i-th row of A, $i \neq j$;*
(III) *multiply the i-th row of A by a nonzero scalar c.*

We shall denote these operations by $I_{(i),(j)}$, $II_{(i)+c(j)}$ and $III_{c(i)}$ respectively. For example, performing $II_{(1)+1(3)}$, $II_{(2)+1(3)}$, $II_{(1)-2(2)}$, $III_{\frac{1}{4}(2)}$, $III_{-1(3)}$, $II_{3+2(2)}$, $I_{(1),(3)}$ in the order given on the matrix

$$A = \begin{bmatrix} 1 & 6 & 3 & 4 \\ 1 & 2 & 1 & 1 \\ -1 & 2 & 1 & 2 \end{bmatrix}$$

produces the matrix

$$\begin{bmatrix} 1 & 0 & 0 & -\frac{1}{2} \\ 0 & 1 & \frac{1}{2} & \frac{3}{4} \\ 0 & 0 & 0 & 0 \end{bmatrix}.$$

Theorem 2.1 *An elementary row operation on a matrix can be achieved by premultiplying the matrix (i.e., multiplying on the left) by an appropriate nonsingular matrix.*

Proof. We shall actually construct the required matrices and exhibit their inverses to show that they are nonsingular. Let $A \in M_{m,n}(R)$ be the given matrix and let $\{e_1, \ldots, e_m\}$ be the standard basis of $M_{1,m}(R)$. That is, e_j is the $1 \times m$ matrix with 1 in position j, zero elsewhere. Then

$$e_i A = A_{(i)}, \quad i = 1, \ldots, m.$$

Let $E_{(i),(j)}$ be the m-square matrix whose ith row is e_j, whose jth row is e_i and whose kth row is e_k, for all k, $1 \leq k \leq m$, different from i and j. Then the ith row of $E_{(i),(j)}A$ is $e_jA = A_{(j)}$; the jth row of $E_{(i),(j)}A$ is $e_iA = A_{(i)}$ and for any other k the kth row of $E_{(i),(j)}A$ is $e_kA = A_{(k)}$. Thus $E_{(i),(j)}A$ is the matrix obtained from A by the elementary operation $I_{(i),(j)}$.

Now, let E_{ij} (without parentheses in the subscripts) denote the m-square matrix defined in Exercise 1 of the preceding section; i.e., E_{ij} is the m-square matrix with 1 in its (i,j) position and 0 elsewhere. Set

$$E_{(i)+c(j)} = I_m + cE_{ij}.$$

Then $E_{(i)+c(j)}A = A + cE_{ij}A$, where $E_{ij}A$ is the $m \times n$ matrix whose ith row is $A_{(j)}$ and whose other rows are all 0. Thus premultiplication by $E_{(i)+c(j)}$ achieves the elementary row operation $II_{(i)+c(j)}$. Lastly set

$$E_{c(i)} = I_m - (1 - c)E_{ii}.$$

Then $E_{c(i)}A = A - (1 - c)E_{ii}A$, where $(1 - c)E_{ii}A$ is the $m \times n$ matrix whose ith row is $1 - c$ times $A_{(i)}$ and whose other rows are 0. Thus premultiplication by $E_{c(i)}$ achieves the elementary row operation $III_{c(i)}$.

Note that, since $E_{(i),(j)} = E_{(i),(j)}I_m$, $E_{(i)+c(j)} = E_{(i)+c(j)}I_m$ and $E_{c(i)} = E_{c(i)}I_m$, the matrices $E_{(i),(j)}$, $E_{(i)+c(j)}$ and $E_{c(i)}$ are obtained by performing the elementary row operations $I_{(i),(j)}$, $II_{(i)+c(j)}$, $III_{c(i)}$ respectively on the identity matrix I_m.

The matrices $E_{(i),(j)}$, $E_{(i)+c(j)}$, $E_{c(i)}$ are nonsingular. In fact,

$$E_{(i),(j)}E_{(j),(i)} = E_{(i),(j)}(E_{(j),(i)}I_m) = I_m,$$
$$E_{(i)+c(j)}E_{(i)-c(j)} = E_{(i)+c(j)}(E_{(i)-c(j)}I_m) = I_m,$$

and

$$E_{c(i)}E_{c^{-1}(i)} = E_{c(i)}(E_{c^{-1}(i)}I_m) = I_m.$$

It follows that

$$E_{(i),(j)}{}^{-1} = E_{(i),(j)}, \tag{5}$$

$$E_{(i)+c(j)}{}^{-1} = E_{(i)-c(j)}, \tag{6}$$

$$E_{c(i)}{}^{-1} = E_{c^{-1}(i)}. \tag{7}$$

The statement (5) says that premultiplication by $E_{(i),(j)}$ interchanges rows i and j; and to undo this operation we switch rows i and j again by another premultiplication by $E_{(i),(j)}$. Similar remarks may be made about statements (6) and (7).

Definition 2.2 (*Elementary matrices*)   *The nonsingular matrices $E_{(i),(j)}$, $E_{(i)+c(j)}$, $E_{c(i)}$ defined in the proof of Theorem 2.1 are called **elementary matrices**.*

It follows immediately from Theorems 1.5 and 2.1 that

Theorem 2.2   *The rank of a matrix is invariant under elementary row operations.*

In the next section we shall use elementary row operations to develop a systematic procedure for solving systems of linear equations and determining the rank of a matrix. Before we do this we shall discuss in this section some other important operations on matrices.

Definition 2.3 (*Elementary column operations*)   *The three elementary column operations on a matrix $A \in M_{m,n}(R)$ are:*

(*I*)   *interchange of the i-th and the j-th column of A;*

(*II*)   *adding c times the j-th column of A to the i-th column of A, $i \neq j$;*

(*III*) *multiplying the i-th column of A by a nonzero scalar c.*

We denote these operations by $\mathrm{I}^{(i),(j)}$, $\mathrm{II}^{(i)+c(j)}$ and $\mathrm{III}^{c(i)}$ respectively.

Theorem 2.3   (a)   *An elementary column operation on a matrix can be achieved by postmultiplying the matrix by an appropriate nonsingular matrix.*

(b)   *The rank of a matrix is invariant under elementary column operations.*

Proof.   Part (a) can be proved in a manner similar to the proof of Theorem 2.1 or it can be deduced from Theorem 2.1 in the following way. Let $A \in M_{m,n}(R)$. Since elementary column operations on A correspond to elementary row operations on A^T, the three elementary column operations can be achieved as follows:

$$I^{(i),(j)}: \qquad (E_{(i),(j)}A^T)^T = AE_{(i),(j)}{}^T = AE_{(i),(j)};$$
$$II^{(i)+c(j)}: \quad (E_{(i)+c(j)}A^T)^T = AE_{(i)+c(j)}{}^T = AE_{(j)+c(i)};$$
$$III^{c(i)}: \qquad (E_{c(i)}A^T)^T = AE_{c(i)}{}^T = AE_{c(i)}.$$

The matrices used in elementary column operations are denoted by $E^{(i),(j)}$, $E^{(i)+c(j)}$, $E^{c(i)}$. Therefore

$$E^{(i),(j)} = E_{(i),(j)}, \tag{8}$$
$$E^{(i)+c(j)} = E_{(j)+c(i)}, \tag{9}$$
$$E^{c(i)} = E_{c(i)} \tag{10}$$

and, by Theorem 2.1, these matrices are nonsingular. It follows that the rank is invariant under elementary column operations. Also, $E^{(i),(j)}$, $E^{(i)+c(j)}$, $E^{c(i)}$ can be obtained by performing $I^{(i),(j)}$, $II^{(i)+c(j)}$, $III^{c(i)}$, respectively, on the columns of I_n.

For example, performing $II^{(2)-6(1)}$, $II^{(3)-3(1)}$, $II^{(4)-4(1)}$, $III^{-\frac{1}{4}(2)}$, $II^{(3)+2(2)}$, $II^{(4)+3(2)}$, in the order given, on the matrix

$$A = \begin{bmatrix} 1 & 6 & 3 & 4 \\ 1 & 2 & 1 & 1 \\ -1 & 2 & 1 & 2 \end{bmatrix}$$

produces the matrix

$$B = \begin{bmatrix} 1 & 0 & 0 & 0 \\ 1 & 1 & 0 & 0 \\ -1 & -2 & 0 & 0 \end{bmatrix}.$$

It follows that

$$B = AE$$

where E is the matrix

$$E^{(2)-6(1)}E^{(3)-3(1)}E^{(4)-4(1)}E^{-\frac{1}{4}(2)}E^{(3)+2(2)}E^{(4)+3(2)}.$$

A straightforward computation yields

$$E = \begin{bmatrix} 1 & \frac{3}{2} & 0 & \frac{1}{2} \\ 0 & -\frac{1}{4} & -\frac{1}{2} & -\frac{3}{4} \\ 0 & 0 & 1 & 0 \\ 0 & 0 & 0 & 1 \end{bmatrix}.$$

Definition 2.4 (*Equivalent matrices*) *If A can be obtained from B by a sequence of elementary row operations (row and column operations) then A and B are called* **row equivalent** (**equivalent**).

An important method, mostly from a theoretical point of view, of performing multiplication of matrices is so-called block multiplication. This type of multiplication is also useful in analyzing various practical schemes for inverting matrices and solving large scale systems of linear equations. We do not go into these applications here.

Definition 2.5 (*Conformal partitioning*) *Let the matrices $A \in M_{m,n}(R)$ and $B \in M_{n,p}(R)$ be partitioned into submatrices $A^{i,j}$ and $B^{i,j}$ respectively as follows:*

$$
A = \begin{array}{c} \\ \\ \end{array}
\begin{array}{cccc} \overset{n_1}{} & \overset{n_2}{} & \cdots & \overset{n_h}{} \end{array}
\left[
\begin{array}{cccc}
\overbrace{A^{1,1}} & \overbrace{A^{1,2}} & \cdots & \overbrace{A^{1,h}} \\
A^{2,1} & A^{2,2} & \cdots & A^{2,h} \\
\cdot & & & \cdot \\
\cdot & & & \cdot \\
\cdot & & & \cdot \\
A^{g,1} & A^{g,2} & \cdots & A^{g,h}
\end{array}
\right]
\begin{array}{l}
\}m_1 \\ \}m_2 \\ \cdot \\ \cdot \\ \cdot \\ \}m_g
\end{array}
$$

where $A^{i,j}$ is $m_i \times n_j$, $\displaystyle\sum_{i=1}^{g} m_i = m$, $\displaystyle\sum_{j=1}^{h} n_j = n$;

$$
B = \begin{array}{c} \\ \\ \end{array}
\begin{array}{cccc} \overset{p_1}{} & \overset{p_2}{} & \cdots & \overset{p_k}{} \end{array}
\left[
\begin{array}{cccc}
\overbrace{B^{1,1}} & \overbrace{B^{1,2}} & \cdots & \overbrace{B^{1,k}} \\
B^{2,1} & B^{2,2} & \cdots & B^{2,k} \\
\cdot & \cdot & & \cdot \\
\cdot & \cdot & & \cdot \\
\cdot & \cdot & & \cdot \\
B^{h,1} & B^{h,2} & \cdots & B^{h,k}
\end{array}
\right]
\begin{array}{l}
\}n_1 \\ \}n_2 \\ \cdot \\ \cdot \\ \cdot \\ \}n_h
\end{array}
$$

where $B^{i,j}$ is $n_i \times p_j$, $\displaystyle\sum_{i=1}^{h} n_i = n$, $\displaystyle\sum_{j=1}^{k} p_j = p$. Then A and B (in this order) are said to be **conformally partitioned** *for multiplication. If the entries of a partitioned matrix A are written out, then the partitioning into blocks $A^{i,j}$ is usually indicated by horizontal and vertical dashed lines.*

Theorem 2.4   *Let $A \in M_{m,n}(R)$ and $B \in M_{n,p}(R)$ be partitioned conformally for multiplication as in Def. 2.5. Let $C = AB$ be partitioned as follows:*

$$
C = \begin{bmatrix}
\overbrace{C^{1,1}}^{p_1} & \overbrace{C^{1,2}}^{p_2} & \cdots & \overbrace{C^{1,k}}^{p_k} \\[-1pt]
C^{2,1} & C^{2,2} & \cdots & C^{2,k} \\
\cdot & \cdot & & \cdot \\
\cdot & \cdot & & \cdot \\
\cdot & \cdot & & \cdot \\
C^{g,1} & C^{g,2} & \cdots & C^{g,k}
\end{bmatrix}
\begin{matrix} \}m_1 \\ \}m_2 \\ \cdot \\ \cdot \\ \cdot \\ \}m_g \end{matrix}
\tag{11}
$$

where $C^{i,j}$ is $m_i \times p_j$, $i = 1, \ldots, g$, $j = 1, \ldots, k$. Then

$$
C^{i,j} = \sum_{t=1}^{h} A^{i,t} B^{t,j}, \quad i = 1, \ldots, g, \quad j = 1, \ldots, k. \tag{12}
$$

Proof.   We shall show that the (r,s) entry in $C^{i,j}$ is equal to the (r,s) entry in

$$
\sum_{t=1}^{h} A^{i,t} B^{t,j}.
$$

Now, $(C^{i,j})_{rs} = c_{r's'}$, where $r' = m_1 + \cdots + m_{i-1} + r$ and $s' = p_1 + \cdots + p_{j-1} + s$. Thus

$$
(C^{i,j})_{rs} = \sum_{t=1}^{h} a_{r't} b_{ts'}.
$$

On the other hand,

$$
\left(\sum_{t=1}^{h} A^{i,t} B^{t,j} \right)_{rs} = \sum_{t=1}^{h} (A^{i,t} B^{t,j})_{rs}
$$

$$
= \sum_{t=1}^{h} \sum_{q=1}^{n_t} (A^{i,t})_{rq} (B^{t,j})_{qs}.
$$

Now

$$
(A^{i,t})_{rq} = a_{r'q'} \quad \text{and} \quad (B^{t,j})_{qs} = b_{q's'}
$$

where $r' = m_1 + \cdots + m_{i-1} + r$, $q' = n_1 + \cdots + n_{t-1} + q$ and $s' = p_1 + \cdots + p_{j-1} + s$.

Therefore

$$\left(\sum_{t=1}^{h} A^{i,t} B^{t,j} \right)_{rs} = \sum_{t=1}^{h} \left(\sum_{q'=n_1+\cdots+n_{t-1}+1}^{n_1+\cdots+n_{t-1}+n_t} a_{r'q'} b_{q's'} \right)$$

$$= \sum_{t=1}^{n} a_{r't} b_{ts'}.$$

We illustrate this with an example. Let $A \in M_{m,n}(R)$ and $B \in M_{n,p}(R)$. Let A be partitioned into 1×1 blocks, i.e.,

$$A^{i,t} = a_{ij}, \quad i = 1, \ldots, m, \quad t = 1, \ldots, n,$$

and B into $1 \times p$ blocks, i.e.,

$$B^{t,j} = B_{(t)}, \quad t = 1, \ldots, n, \quad \text{and} \quad j = 1.$$

Then

$$(AB)_{(i)} = (AB)^{i,j} = \sum_{t=1}^{n} a_{it} B_{(t)}, \quad i = 1, \ldots, m, \quad j = 1,$$

as in Theorem 1.3.

Definition 2.6 (Direct sum of matrices) *If $A \in M_n(R)$ can be partitioned into blocks so that the blocks $A^{i,i}$ are $n_i \times n_i$, $i = 1, \ldots, h$, and $A^{i,j} = 0$ for $i \neq j$, then A is said to be the direct sum of $A^{1,1}, \ldots, A^{h,h}$ and this fact is denoted by*

$$A = A^{1,1} \dotplus \cdots \dotplus A^{h,h}$$

or

$$A = \sum_{i=1}^{h} {}^{\cdot} A^{i,i}.$$

The direct sum of matrices can also be defined independently as follows. If $B = (b_{ij}) \in M_p(R)$ and $C = (c_{ij}) \in M_q(R)$, then

$$S = (s_{ij}) = B \dotplus C$$

if

$$s_{ij} = \begin{cases} b_{ij}, & i,j = 1, \ldots, p, \\ c_{i-p,j-p}, & i,j = p + 1, \ldots, p + q, \\ 0, & \text{otherwise.} \end{cases}$$

Further, if $A_j \in M_{n_j}(R), j = 1, \ldots, k$, then

$$\sum_{j=1}^{k}{}^{\cdot} A_j = \sum_{j=1}^{k-1}{}^{\cdot} A_j \dotplus A_k.$$

The concept of direct sum is of paramount importance in many parts of matrix theory.

Theorem 2.5   *If $A_j \in M_{n_j}(R)$ and $B_j \in M_{n_j}(R), j = 1, \ldots, k$, then*

$$\sum_{j=1}^{k}{}^{\cdot} A_j + \sum_{j=1}^{k}{}^{\cdot} B_j = \sum_{j=1}^{k}{}^{\cdot} (A_j + B_j), \tag{13}$$

$$\left(\sum_{j=1}^{k}{}^{\cdot} A_j \right) \left(\sum_{j=1}^{k}{}^{\cdot} B_j \right) = \sum_{j=1}^{k}{}^{\cdot} A_j B_j, \tag{14}$$

$$\rho \left(\sum_{j=1}^{k}{}^{\cdot} A_j \right) = \sum_{j=1}^{k} \rho(A_j), \tag{15}$$

where $\rho(X)$ is the rank of the matrix X.

Proof.   Formula (13) follows from Def. 2.6 and the definition of matrix addition. Formula (14) follows from Def. 2.6 and Theorem 2.4. To prove (15), let

$$B = \sum_{j=1}^{k}{}^{\cdot} A_j$$

and define

$$m_j = \sum_{i=1}^{j-1} n_i, \quad j = 2, \ldots, k, \quad m_1 = 0.$$

Then if $R(X)$ again denotes the row space of the matrix X, we have

$$R\left(\sum_{j=1}^{k} A_j\right) = \sum_{j=1}^{k} \langle B_{(m_j+1)}, \ldots, B_{(m_j+n_j)}\rangle,$$

$$\langle B_{(m_i+1)}, \ldots, B_{(m_i+n_i)}\rangle \cap \langle B_{(m_j+1)}, \ldots, B_{(m_j+n_j)}\rangle = 0$$

for $i \neq j$ and

$$\rho(A_j) = \dim \langle B_{(m_j+1)}, \ldots, B_{(m_j+n_j)}\rangle.$$

Therefore

$$\rho\left(\sum_{j=1}^{k} A_j\right) = \dim\left(\sum_{j=1}^{k} \langle B_{(m_j+1)}, \ldots, B_{(m_j+n_j)}\rangle\right)$$

$$= \sum_{j=1}^{k} \dim \langle B_{(m_j+1)}, \ldots, B_{(m_j+n_j)}\rangle$$

$$= \sum_{j=1}^{k} \rho(A_j),$$

the second equality holding by virtue of Theorem 3.3, Chap. 1.

Quiz

Answer true or false:

1. If a row of a matrix A is multiplied by any scalar, then the rank of the resulting matrix is equal to $\rho(A)$.
2. If $A \in M_{m,n}(R)$, then

$$\langle A_{(1)} + A_{(2)}, A_{(1)}, A_{(3)}, \ldots, A_{(m)}\rangle = \langle A_{(1)}, \ldots, A_{(m)}\rangle.$$

3. If $A \in M_{m,n}(R)$, then

$$\langle A_{(1)} + A_{(2)}, A_{(1)} + A_{(2)}, A_{(3)}, \ldots, A_{(m)}\rangle = \langle A_{(1)}, \ldots, A_{(m)}\rangle.$$

4. If $A \in M_{m,n}(R)$, then

$$\langle A_{(1)} + A_{(2)}, A_{(1)} - A_{(2)}, A_{(3)}, \ldots, A_{(m)}\rangle = \langle A_{(1)}, \ldots, A_{(m)}\rangle.$$

5. If $c_j \in R, j = 1, \ldots, n$, then

$$\operatorname{diag}(c_1, \ldots, c_n) = c_1 I_1 \dotplus \cdots \dotplus c_n I_1.$$

6. The matrix $I_m - E_{ii} - E_{jj} + E_{ij} + E_{ji}$ is an elementary matrix of type I.

7. If $E_{(i),(j)}$ is an elementary n-square matrix of type I, then

$$E_{(i),(j)} = E_{(j),(i)} = E^{(i),(j)} = E_{(i),(j)}{}^{-1}.$$

8. If $E^{(i) + c(j)}$ is an n-square elementary matrix of type II then

$$(E^{(i) + c(j)})^{-1} = E_{(j) + c(i)}.$$

9. Let A be the partitioned matrix $\begin{bmatrix} B & C \\ \hline D & E \end{bmatrix}$ where B, C, D, E are nonsingular matrices. Then

$$A^{-1} = \begin{bmatrix} B^{-1} & C^{-1} \\ \hline D^{-1} & E^{-1} \end{bmatrix}.$$

10. If A is the partitioned matrix $\begin{bmatrix} B & C \\ \hline D & E \end{bmatrix}$ and

$$\rho(B) = \rho(C) = \rho(D) = \rho(E) = m, \text{ then } \rho(A) = 2m.$$

Exercises

1. Apply, in the order given, the elementary row operations $I_{(1),(3)}$, $II_{(2)-1(1)}$, $II_{(3)-1(2)}$, $II_{(2)-2(1)}$, $III_{\frac{1}{10}(2)}$ to the matrix

$$A = \begin{bmatrix} 2 & 4 & 2 \\ 3 & 1 & 1 \\ 1 & -3 & -1 \end{bmatrix}.$$

2. Denote the matrix obtained in Exercise 1 by B. Find a matrix E such that $EA = B$. Compute E^{-1}.

3. Let

$$C = \begin{bmatrix} 1 & 1 & 1 \\ 3 & 1 & 2 \\ -1 & -1 & 1 \end{bmatrix}.$$

Show that C is row equivalent to I_3. Hence find C^{-1}.

4. Show that if A_i, $i = 1, \ldots, k$, are nonsingular matrices, then

$$\left(\sum_{i=1}^{k} A_i \right)^{-1} = \sum_{i=1}^{k} A_i^{-1}.$$

5. Let $B \in M_k(R)$ and $C \in M_{n-k}(R)$. Let T be the linear transformation of $V_n(R)$ into itself defined by

$$T(x) = (B + C)x, \quad x \in V_n(R).$$

Find a nonzero subspace U of $V_n(R)$ invariant under T, i.e., such that

$$T(u) \in U \text{ for all } u \in U.$$

2.3 Linear Equations

Elementary row operations allow us to reduce a matrix to a simple standard form, the so-called *Hermite normal form*. This reduction of a matrix A provides a method for finding all solutions of a system of linear equations $Ax = b$.

Theorem 3.1   (*Hermite normal form*)   *Let $A \in M_{m,n}(R)$ be a matrix of rank $\rho(A) = r$. Then A is row equivalent to a matrix $B \in M_{m,n}(R)$ of the following form:*
  (i) $B_{(i)} \neq 0$ for $i = 1, \ldots, r$; $B_{(i)} = 0, i = r+1, \ldots, m$;
  (ii) *there is a sequence of integers $n_1, \ldots, n_r, 1 \leq n_1 < n_2 < \cdots < n_r \leq n$, such that*

$$b_{ij} = 0 \text{ for } j < n_i, \quad i = 1, \ldots, r,$$

and

$$B^{(n_i)} = e_i, \quad i = 1, \ldots, r,$$

where $\{e_1, \ldots, e_m\}$ is the standard basis of $V_m(R)$; i.e., B has the form

$$
B = \begin{bmatrix}
0 \cdots 0 \ 1 \ x & \cdots \ x \ 0 \ x & \cdots \ x \ 0 \ x & \cdots & 0 & \cdots \ x \\
0 \cdots & 0 & \cdots \ 0 \ 1 \ x & \cdots \ x \ 0 \ x & \cdots & 0 & \cdots \ x \\
0 \cdots & 0 & \cdots & 0 & \cdots \ 0 \ 1 \ x & \cdots & 0 & \cdots \ x \\
\cdots & \cdots & \cdots & \cdots & \cdots & \cdots \\
0 \cdots & 0 & \cdots & 0 & \cdots & 0 & \cdots \ 0 \ 1 \ x \ \cdots \ x \\
0 \cdots & & \cdots & & \cdots & & \cdots \ 0 \\
\cdots & \cdots & \cdots & \cdots & \cdots \\
0 \cdots & & \cdots & & \cdots & & \cdots \ 0
\end{bmatrix} \quad (1)
$$

where x represents unspecified entries.

Proof.   We use induction on m. If $A = 0$, then A is already in Hermite normal form. Otherwise, let n_1 be the first nonzero column of A. By a type I elementary row operation bring a nonzero entry into the $(1, n_1)$ position and by a type III elementary row operation make this entry 1. Use type II row operations to make all other entries in the column n_1 equal to 0. Thus A is row equivalent to a matrix

$$C = \left[\begin{array}{ccccccc} 0 & \cdots & 0 & 1 & x & \cdots & x \\ \hline & & & C' & & & \end{array}\right]$$

where C' is a $(m - 1) \times n$ submatrix of C such that $(C')^{(j)} = 0, j = 1,$ $\ldots$, n_1. Therefore in the n_1th column of C the only nonzero entry is in the first row. Thus

$$C_{(1)} \notin R(C') \quad \text{and} \quad \rho(C') = r - 1.$$

Also, by the induction hypothesis, C' is row equivalent to a matrix B' in Hermite normal form:

(i) $B'_{(i)} \neq 0$ for $i = 1, \ldots , r - 1$; $B'_{(i)} = 0$ for $i = r, \ldots ,$ $m - 1$;

(ii) $(B')_{ij} = 0$ for $j < n_i, i = 2, \ldots , r$, and

$$B^{(n_i)} = e'_{i - 1}, \quad i = 2, \ldots , r, \quad \text{where} \quad \{e'_1, \ldots , e'_{m - 1}\}$$

is the standard basis of $V_{m - 1}(R)$. Clearly no row operations can affect the first n_1 columns of C' since they are all zero. Now, by elementary row operations of type II on the matrix

$$\left[\begin{array}{ccccccc} 0 & \cdots & 0 & 1 & x & \cdots & x \\ \hline & & & B' & & & \end{array}\right]$$

make each of the entries $(1, n_2), (1, n_3), \ldots , (1, n_r)$ equal to 0. The resulting matrix B is in Hermite normal form, is row equivalent to C, and is therefore row equivalent to A. Clearly the number of nonzero rows in B is equal to r. We show that the first r rows of B are linearly independent. For,

$$\sum_{i = 1}^{n} d_i B_{(i)} = 0 \quad \text{implies} \quad \sum_{i = 1}^{r} d_i B_{i, n_t}, t = 1, \ldots , r,$$

and, since $B_{i, n_t} = \delta_{it}$, we conclude that $d_i = 0, t = 1, \ldots , r$. Thus $\rho(B) = r = \rho(A)$.

The proof of Theorem 3.1 is constructive, as we shall demonstrate in the following example.

Let

$$A = \begin{bmatrix} 0 & 0 & 0 & 0 & 1 & 1 & 1 \\ 0 & 2 & 6 & 2 & 0 & 0 & 4 \\ 0 & 1 & 3 & 1 & 1 & 0 & 1 \\ 0 & 1 & 3 & 1 & 2 & 1 & 2 \end{bmatrix}.$$

The elementary row operations $I_{(1),(3)}$, $II_{(2)-2(1)}$ and $II_{(4)-1(1)}$ transform A into the row equivalent matrix

$$C = \begin{bmatrix} 0 & 1 & 3 & 1 & 1 & 0 & 1 \\ 0 & 0 & 0 & 0 & -2 & 0 & 2 \\ 0 & 0 & 0 & 0 & 1 & 1 & 1 \\ 0 & 0 & 0 & 0 & 1 & 1 & 1 \end{bmatrix}.$$

We concentrate our attention now on the submatrix C' of C lying in rows 2, 3, 4 and columns 1, 2, 3, 4, 5, 6, 7 of C. By the elementary operations $III_{-\frac{1}{2}(2)}$, $II_{(3)-1(2)}$, $II_{(4)-1(2)}$, $II_{(4)-1(3)}$ on rows 2, 3 and 4 of C we reduce C' to its Hermite normal form. Finally, $II_{(1)-1(2)}$ produces

$$B = \begin{bmatrix} 0 & 1 & 3 & 1 & 0 & 0 & 2 \\ 0 & 0 & 0 & 0 & 1 & 0 & -1 \\ 0 & 0 & 0 & 0 & 0 & 1 & 2 \\ 0 & 0 & 0 & 0 & 0 & 0 & 0 \end{bmatrix},$$

the Hermite normal form of A.

In this case, $n_1 = 2$, $n_2 = 5$, $n_3 = 6$, $r = 3$. Note that

$$B = EA,$$

where

$$E = E_{(1)-1(2)}E_{(4)-1(3)}E_{(4)-1(2)}E_{(3)-1(2)}E_{-\frac{1}{2}(2)}E_{(4)-1(1)}E_{(2)-2(1)}E_{(1),(3)}$$

$$= \begin{bmatrix} 0 & \frac{1}{2} & 0 & 1 \\ 0 & -\frac{1}{2} & 1 & 0 \\ 1 & \frac{1}{2} & -1 & 0 \\ -1 & 0 & -1 & 1 \end{bmatrix}.$$

Theorem 3.2

 (a)   *A matrix is nonsingular if and only if it is row equivalent to an identity matrix.*

 (b)   *An $n \times n$ matrix A is nonsingular if and only if $n = \rho(A)$.*

 (c)   *A matrix is nonsingular if and only if it is a product of elementary matrices.*

Proof. Suppose A is nonsingular so that $AA^{-1} = A^{-1}A = I_n$. Let $B = E_k \ldots E_1 A$, where $E_1, \ldots , E_k$ are elementary matrices, and B is the Hermite normal form of A. Now, by Theorem 2.1, $E_1, \ldots , E_k$ are nonsingular and therefore $B^{-1} = A^{-1}E_1^{-1} \ldots E_k^{-1}$ and B is nonsingular. Thus B cannot have a zero row, otherwise BB^{-1} would have a zero row. Therefore $n = \rho(B) = \rho(A)$ and clearly the integers $n_1 < \ldots < n_r$ in Theorem 3.1(ii) must be $1, 2, \ldots , n$. Since $B^{(n_i)} = e_i$, $i = 1, \ldots , n$, it follows that $B = I_n$. Hence

$$E_k \cdots E_1 A = I_n,$$

and A is row equivalent to I_n. Moreover, it follows that

$$A = E_1^{-1} \cdots E_k^{-1}$$

and, by (5), (6), (7) in Sec. 2, A is a product of elementary matrices. We have proved that if A is nonsingular then it is row equivalent to I_n, $\rho(A) = n$, and A is a product of elementary matrices.

Conversely, if A is row equivalent to the identity matrix, then

$$E_k \cdots E_1 A = I_n$$

for some elementary, and therefore nonsingular, matrices $E_1, \ldots , E_k$. Thus,

$$A^{-1} = E_k \cdots E_1$$

and A is nonsingular. If $\rho(A) = n$, then the Hermite normal form of A is I_n,

$$E_k \cdots E_1 A = I_n,$$

and again A is nonsingular.

Finally, if A is a product of elementary matrices it is obviously non-singular as before.

We are now in a position to describe a systematic procedure for solving the system of linear equations

$$Ax = b \tag{2}$$

where $A \in M_{m,n}(R)$ is of rank r, $x = (x_1, \ldots, x_n)$, and $b = (b_1, \ldots, b_m)$. Let $B = (b_{ij}) = EA$ be the Hermite normal form of A. Then the system (2) is equivalent to

$$EAx = Eb \tag{3}$$

or

$$Bx = c \tag{4}$$

where $c = (c_1, \ldots, c_m) = Eb$. The system (4) can also be written in the form

$$B_{(i)}x = c_i, \quad i = 1, \ldots, m. \tag{5}$$

Thus (4) has solutions only if $c_i = 0$ for $i = r + 1, \ldots, m$. If this is the case then (4), and therefore (2), always has solutions. For, taking into account that B is in Hermite normal form, (4) can be written as

$$x_{n_i} + \sum_{j \notin N} b_{ij}x_j = c_i, \quad i = 1, \ldots, r,$$

where $N = \{n_1, \ldots, n_r\}$, or

$$x_{n_i} = c_i - \sum_{j \notin N} b_{ij}x_j, \quad i = 1, \ldots, r. \tag{6}$$

(The symbol $\sum_{j \notin N} b_{ij}x_j$ means the sum of the terms $b_{ij}x_j$ for $j \notin N$.)

Thus all solutions of (2) can be obtained by assigning arbitrary values in R to x_j, $j \notin N$, and then computing the corresponding values for

x_{n_i}, $i = 1, \ldots, r$. In a vector form (6) can be written as follows:

$$x = (x_1, \ldots, x_{n_1}, \ldots, x_{n_2}, \ldots, x_{n_r}, \ldots, x_n) \qquad (7)$$
$$= (x_1, \ldots, c_1 - \sum_{j \notin N} b_{1j}x_j, \ldots, c_2 - \sum_{j \notin N} b_{2j}x_j, \ldots,$$
$$c_r - \sum_{j \notin N} b_{rj}x_j, \ldots, x_n)$$
$$= c' + \sum_{j \notin N} x_j u_j$$

where

$$u_j = e_j - \sum_{i=1}^{r} b_{ij}e_{n_i}, \ j \notin N, \quad \text{and} \quad c' = \sum_{i=1}^{r} c_i e_{n_i}.$$

Setting $x_j = 0$, $j \notin N$, in (6) we see that c' is a particular solution of (2). Moreover the $n - r$ vectors u_j, $j \notin N$, form a basis for the null space of B and hence of A. For, if we set $b = 0$ in (2), then $c = 0$ and all solutions of the system

$$Ax = 0$$

are given by (7) as all linear combinations of the u_j, $j \notin N$. If $j \notin N$ then u_j has a 1 in position j and u_i has a zero in position j for all $i \notin N$, $i \neq j$. Hence u_j, $j \notin N$, are linearly independent.

We illustrate the preceding discussion by solving the system

$$Ax = b$$

where A is the matrix used in the example which follows the proof of Theorem 3.1 and $b = (-4, 2, 2, -2)$. Then as before,

$$B = \begin{bmatrix} 0 & 1 & 3 & 1 & 0 & 0 & 2 \\ 0 & 0 & 0 & 0 & 1 & 0 & -1 \\ 0 & 0 & 0 & 0 & 0 & 1 & 2 \\ 0 & 0 & 0 & 0 & 0 & 0 & 0 \end{bmatrix}$$

and

$$E = \begin{bmatrix} 0 & \frac{1}{2} & 0 & 0 \\ 0 & -\frac{1}{2} & 1 & 0 \\ 1 & \frac{1}{2} & -1 & 0 \\ -1 & 0 & -1 & 1 \end{bmatrix}.$$

Therefore

$$c = (1,1,-5,0).$$

Now $N = \{2,5,6\}$ and we compute

$$u_1 = e_1 = (1,0,0,0,0,0,0),$$
$$u_3 = e_3 - 3e_2 = (0,-3,1,0,0,0,0),$$
$$u_4 = e_4 - e_2 = (0,-1,0,1,0,0,0),$$
$$u_7 = e_7 - 2e_2 + e_5 - 2e_6 = (0,-2,0,0,1,-2,1),$$
$$c' = e_2 + e_5 - 5e_6 = (0,1,0,0,1,-5,0).$$

We verify that $Au_j = 0$, $j = 1,3,4,7$, and since $r = \rho(A) = \rho(B) = 3$, we know that the vectors u_1,u_3,u_4 and u_7 (i.e., u_j, $j \notin N$) form a basis for the null space of A. Every solution x of the system $Ax = b$ is obtained by assigning values to the scalars x_1,x_3,x_4,x_7 in

$$x = c' + x_1u_1 + x_3u_3 + x_4u_4 + x_7u_7.$$

In the following theorem we put together some results on systems of linear equations most of which we proved before, though perhaps in a somewhat different guise.

Theorem 3.3 (a) *The solutions x of the homogeneous system of linear equations*

$$Ax = 0, \quad A \in M_{m,n}(R),\tag{8}$$

form a space of dimension $n - \rho(A)$. In particular, if A is nonsingular ($m = n$) the only solution of (8) is the trivial one, $x = 0$.
  (b) *If u is a solution of (8) and v is a solution of*

$$Ax = b, \quad b \in V_m(R),\tag{9}$$

then $u + v$ is a solution of (9). Moreover if v is a particular solution of (9) then any solution of (9) is of the form $u + v$ where u is a solution of (8).
  (c) *The system (9) has a solution if and only if $b \in \langle A^{(1)}, \ldots , A^{(n)}\rangle$.*
  (d) *If $A \in M_n(R)$, then (9) has a unique solution if and only if A is nonsingular.*

Proof. Part (a) is merely a restatement of Theorem 2.1 (i), Chap. 1. Parts (b) and (c) were essentially proved in an earlier part of this section. It is easy, however, to prove them independently. For, if

$$Au = 0 \quad \text{and} \quad Av = b$$

then

$$A(u + v) = Au + Av = 0 + b = b.$$

Also, if $Aw = b$, then $u = w - v$ is a solution of (8), and $w = u + v$. Part (c) is obvious, since (9) can be written in the form

$$\sum_{j=1}^{n} A^{(j)}x_j = b$$

where $x = (x_1, \ldots, x_n)$ (see also Theorem 2.3, Chap. 1). Finally, part (d) follows immediately from the preceding parts.

We conclude with a theorem on equivalence (see Def. 2.4).

Theorem 3.4   (a)   *Every $m \times n$ matrix A is equivalent to a matrix $C = (c_{ij})$ where $c_{ii} = 1$, $i = 1, \ldots, \rho(A)$, and $c_{ij} = 0$ otherwise. The matrix C is called the canonical form of A.*

(b)   *Two $m \times n$ matrices are equivalent if and only if their ranks are equal.*

Proof.   Let B be the Hermite normal form of A, and let $\rho(A) = r$. On the matrix B perform a sequence of elementary column operations of type II to make every entry in the first row, except the $(1, n_1)$ entry, equal to 0. Next by type II elementary column operations on the matrix now at hand reduce every entry in the second row, except the $(2, n_2)$ entry, to 0; and so on. Finally perform the operations $I^{(1),(n_1)}$, $I^{(2),(n_2)}, \ldots, I^{(r),(n_r)}$ in order. The resulting matrix is C.

To prove part (b), note that all $m \times n$ matrices of rank r are equivalent to the same canonical form. We leave the formal justification of this statement as an exercise for the student. Conversely, by Theorem 1.5, equivalent matrices have the same rank.

Quiz

Answer true or false:

1.   A system of m equations in $m + 1$ unknowns always has a solution.

2. The set of all solution vectors of the system

$$Ax = b$$

forms a vector space.
3. Any two nonsingular n-square matrices are equivalent.
4. Any n-square matrix A in Hermite normal form is upper triangular (i.e., the elements a_{ij}, $i > j$, are zero).
5. The Hermite normal form of A^{-1} is equal to the inverse of the Hermite normal form of A.
6. The system $Ax = b$, where A is a $m \times n$ matrix of rank m, always has a solution.
7. The Hermite normal form of A is a zero $m \times n$ matrix if and only if $A = O_{m,n}$.
8. If A^{-1} and B^{-1} are both n-square, they then have the same Hermite normal form.
9. If A is nonsingular, then $A^{-1}b$ is the only solution of $Ax = b$.
10. A system

$$\sum_{j=1}^{n} a_{ij}x_j = 0, \quad i = 1, \ldots, m,$$

has a nonzero solution if and only if $n > m$.

Exercises

1. Write out a detailed proof of Theorem 3.4(b).
2. Express the elementary operation $I_{(i),(j)}$ as a sequence of elementary row operations of types II and III.
3. Use Theorem 3.3(a) to prove that if A is singular then there exists $B \neq 0$ such that $AB = 0$.
4. Completely solve the system

$$\begin{bmatrix} 2 & 1 & 0 & 1 & 3 \\ 1 & 1 & 1 & 1 & 1 \\ 2 & 1 & 1 & 2 & 1 \\ 2 & 2 & 1 & 1 & 4 \end{bmatrix} x = \begin{bmatrix} 3 \\ 3 \\ 3 \\ 6 \end{bmatrix}$$

5. Write the matrix

$$A = \begin{bmatrix} 2 & 1 & 1 & 1 \\ 1 & 2 & 1 & 1 \\ 1 & 1 & 2 & 1 \\ 1 & 1 & 1 & 2 \end{bmatrix}$$

as a product of elementary matrices. Hence find A^{-1}.

2.4 Permutations

We pause in this section to discuss a concept of paramount importance in many parts of algebra and combinatorics: permutations. We shall restrict ourselves to a few theorems on permutations that are required in the theory of determinants.

Definition 4.1 (*Permutation*) *A* **permutation** *on* n *objects, labeled* $1, \ldots , n$, *is a one-one mapping of the set* $\{1, \ldots , n\}$ *onto itself.*

We denote the image of i *under a permutation* σ *by* $\sigma(i)$. *Sometimes it is convenient to denote a permutation as a* $2 \times n$ *array:*

$$\sigma = \begin{pmatrix} 1 & 2 & \cdots & n \\ \sigma(1) & \sigma(2) & \cdots & \sigma(n) \end{pmatrix}. \tag{1}$$

For example,

$$\sigma = \begin{pmatrix} 1 & 2 & 3 & 4 \\ 3 & 2 & 4 & 1 \end{pmatrix}$$

means that σ is the permutation on $\{1,2,3,4\}$ defined by $\sigma(1) = 3$, $\sigma(2) = 2$, $\sigma(3) = 4$ and $\sigma(4) = 1$.

Clearly, the notation (1) is merely a convenient way to exhibit the images of $1, \ldots , n$ under σ and the order of columns is quite immaterial.

Definition 4.2 (*Symmetric group*) *The* **product** *of two permutations* σ *and* φ *is defined in terms of function composition:*

$$(\sigma\varphi)(i) = \sigma[\varphi(i)], \quad i = 1, \ldots , n.$$

The set of all permutations on n *objects together with this operation of multiplication is called the* **symmetric group of degree** n *and is denoted by* S_n.

Before we list some properties of the symmetric group S_n (some of which will merely justify our use of the term "group"), we shall give a few examples of permutation multiplication:

(i) $\begin{pmatrix} 1 & 2 & 3 & 4 & 5 \\ 2 & 4 & 1 & 5 & 3 \end{pmatrix} \begin{pmatrix} 1 & 2 & 3 & 4 & 5 \\ 4 & 1 & 2 & 5 & 3 \end{pmatrix} = \begin{pmatrix} 1 & 2 & 3 & 4 & 5 \\ 5 & 2 & 4 & 3 & 1 \end{pmatrix};$

(ii) $\begin{pmatrix} 1 & 2 & 3 & 4 & 5 \\ 4 & 1 & 2 & 5 & 3 \end{pmatrix} \begin{pmatrix} 4 & 1 & 2 & 5 & 3 \\ 1 & 2 & 3 & 4 & 5 \end{pmatrix} = \begin{pmatrix} 1 & 2 & 3 & 4 & 5 \\ 1 & 2 & 3 & 4 & 5 \end{pmatrix};$

(iii) $\begin{pmatrix} 1 & 2 & 3 & 4 & 5 \\ 5 & 2 & 4 & 3 & 1 \end{pmatrix} \begin{pmatrix} 4 & 1 & 2 & 5 & 3 \\ 1 & 2 & 3 & 4 & 5 \end{pmatrix} = \begin{pmatrix} 1 & 2 & 3 & 4 & 5 \\ 2 & 4 & 1 & 5 & 3 \end{pmatrix};$

(iv) $\begin{pmatrix} 1 & 2 & 3 & 4 & 5 \\ 4 & 1 & 2 & 5 & 3 \end{pmatrix} \begin{pmatrix} 1 & 2 & 3 & 4 & 5 \\ 2 & 4 & 1 & 5 & 3 \end{pmatrix} = \begin{pmatrix} 1 & 2 & 3 & 4 & 5 \\ 1 & 5 & 4 & 3 & 2 \end{pmatrix}.$

For example, in (i), 3 is mapped into 2 by the factor on the right and then 2 is mapped into 4 by the other factor. Thus the product maps 3 into 4 as indicated. Examples (i) and (iv) show that multiplication of permutations is not always commutative, i.e., $\sigma\varphi \neq \varphi\sigma$, in general.

Theorem 4.1 (a) *A product of two permutations in S_n is a permutation in S_n.*
  (b) *Multiplication of permutations is associative.*
  (c) *If*

$$e = \begin{pmatrix} 1 & 2 & \cdots & n \\ 1 & 2 & \cdots & n \end{pmatrix}$$

  and σ is any other permutation in S_n, then $e\sigma = \sigma e = \sigma$.
  (d) *For every $\sigma \in S_n$ there exists $\sigma^{-1} \in S_n$ such that $\sigma\sigma^{-1} = \sigma^{-1}\sigma = e$.*
  (e) *There are $n!$ distinct permutations in S_n.*
  (f) *If $S_n = \{\sigma_1, \sigma_2, \ldots, \sigma_{n!}\}$ and φ is a fixed permutation in S_n then*

$$\{\varphi\sigma_1, \varphi\sigma_2, \ldots, \varphi\sigma_{n!}\} = \{\sigma_1\varphi, \sigma_2\varphi, \ldots, \sigma_{n!}\varphi\} = S_n.$$

Proof. Parts (a) and (c) follow immediately from Def. 4.2. The statement in part (b) is true for function composition in general. If σ, φ, $\tau \in S_n$, then using the definition of multiplication, we have

$$[(\sigma\varphi)\tau](i) = (\sigma\varphi)[\tau(i)] = \sigma\{\varphi[\tau(i)]\},$$

and

$$[\sigma(\varphi\tau)](i) = \sigma[(\varphi\tau)(i)] = \sigma\{\varphi[\tau(i)]\},$$

for any i. Because the multiplication in S_n is associative we can write the product of several permutations without brackets and, in particular, define the kth power of σ by

$$\sigma^1 = \sigma \quad \text{and} \quad \sigma^k = \sigma(\sigma^{k-1}) \quad \text{for } k > 1.$$

To prove (d), set

$$\sigma^{-1} = \begin{pmatrix} \sigma(1) & \sigma(2) & \cdots & \sigma(n) \\ 1 & 2 & \cdots & n \end{pmatrix}$$

Then $\sigma\sigma^{-1} = \sigma^{-1}\sigma = e$.

We prove part (e) by actually counting the number of ways in which the second line of the symbol

$$\begin{pmatrix} 1 & 2 & \cdots & n \\ \sigma(1) & \sigma(2) & \cdots & \sigma(n) \end{pmatrix}$$

can be written. Any of the n numbers $1, \ldots, n$ can be $\sigma(1)$. Once $\sigma(1)$ is chosen, we can choose $\sigma(2)$ to be any of the remaining $n-1$ numbers. After $\sigma(1)$ and $\sigma(2)$ are chosen, we can take for $\sigma(3)$ any of the remaining $n-2$ numbers and so on. The total number of choices is $n(n-1)(n-2) \cdots 2 \cdot 1 = n!$ Hence the number of permutations in S_n is $n!$

To prove that $\{\varphi\sigma_1, \varphi\sigma_2, \ldots, \varphi\sigma_{n!}\} = S_n$ we first show that if $\sigma_i \neq \sigma_j$ then $\varphi\sigma_i \neq \varphi\sigma_j$. For, by part (d), there exists φ^{-1} such that $\varphi^{-1}\varphi = e$ and therefore $\varphi\sigma_i = \varphi\sigma_j$ implies $\varphi^{-1}\varphi\sigma_i = \varphi^{-1}\varphi\sigma_j$ and $\sigma_i = \sigma_j$. Hence all permutations in $\{\varphi\sigma_1, \varphi\sigma_2, \ldots, \varphi\sigma_{n!}\}$ are distinct and, since there are $n!$ of them, they must be, by part (e), the permutations of S_n in some order. Similarly,

$$\{\sigma_1\varphi, \ldots, \sigma_{n!}\varphi\} = S_n.$$

Definition 4.3 *(Cycles)* *A permutation $\sigma \in S_n$ is called a k-cycle if for some $i_1 \in N = \{1, \ldots, n\}$*

$$\sigma^k(i_1) = i_1,$$
$$\sigma^t(i_1) \neq i_1 \quad \text{for } t = 1, \ldots, k-1,$$

and

$$\sigma(j) = j \quad \text{for } j \notin \{i_1, \sigma(i_1), \ldots, \sigma^{k-1}(i_1)\}.$$

We denote σ by $(i_1, \sigma(i_1), \sigma^2(i_1), \ldots, \sigma^{k-1}(i_1))$. In other words, σ is a k-cycle if there exists a subset $K = \{i_1, i_2, \ldots, i_k\}$ of N such that

$$\sigma(i_t) = i_{t+1}, \quad t = 1, \ldots, k-1,$$
$$\sigma(i_k) = i_1,$$

and

$$\sigma(j) = j \quad \text{for } j \notin K.$$

Thus our notation for σ is $\sigma = (i_1, i_2, \ldots, i_k)$. A set of cycles $\sigma_t = (i_{t1}, \ldots, i_{tk_t})$, $t = 1, 2 \ldots$, is said to be *disjoint* if $i_{tr} \neq i_{sm}$ whenever $t \neq s$. As an example of the cycle notation,

$$\begin{pmatrix} 1 & 2 & 3 & 4 & 5 & 6 & 7 & 8 \\ 1 & 4 & 5 & 7 & 2 & 6 & 3 & 8 \end{pmatrix} = (2 \quad 4 \quad 7 \quad 3 \quad 5).$$

Clearly, a 1-cycle is the identity permutation. We also observe that

$$(i_1, i_2, \ldots, i_k) = (i_2, i_3, \ldots, i_1) = \cdots = (i_k, i_1, \ldots, i_{k-1}).$$

The simplest nontrivial cycles, the 2-cycles, play a special part in the theory of permutations. A 2-cycle is called a *transposition*.

Definition 4.4 (*Orbit*) *Let $i \in \{1, \ldots, n\}$ and let $\sigma \in S_n$. The set $\sigma[i] = \{\sigma(i), \sigma^2(i), \ldots, \sigma^k(i)\}$, where k is the least positive integer such that $\sigma^k(i) = i$, is called the **orbit** of i **under** σ. Observe that if t is any positive integer, then $\sigma^t(i)$ is in the orbit of i. For, let $t = kq + r$, where $0 \leq r < k$. Then $\sigma^t(i) = \sigma^{kq+r}(i) = \sigma^r(\sigma^{kq}(i)) = \sigma^r(i)$. Moreover, the orbit of i has exactly k distinct numbers. For, if p and q, $p < q$, are any two numbers in $\{1, \ldots, k\}$ then $\sigma^p(i) = \sigma^q(i)$ implies $\sigma^{q-p}(i) = i$ and, since $0 < q - p < k$, this contradicts our assumption on the minimality of k. The number k is called the **length** of the orbit.*

Theorem 4.2 (a) *The orbits under a permutation $\sigma \in S_n$ divide the set $N = \{1, \ldots, n\}$ into disjoint subsets whose union is N.*

(b) *Every permutation is a product of disjoint cycles. This decomposition into disjoint cycles is unique except for 1-cycles and for the order of the cycles.*

(c) *Every permutation is a product of transpositions.*

Proof.   (a)   Since each number in N belongs to at least one orbit, it suffices to prove that if two orbits have a number in common then they are equal. Suppose then that $\sigma^s(i) = \sigma^t(j)$. We show that i belongs to the orbit of j, and vice versa. Let k be the length of the orbit of i. If $k = s$ there is nothing to prove. If $s < k$ then

$$\sigma^{k-s}(\sigma^s(i)) = \sigma^{k-s}(\sigma^t(j)),$$

or

$$i = \sigma^{k-s+t}(j).$$

Similarly, j belongs to the orbit of i.

(b)   Let $\sigma \in S_n$. Let $i_1 \in N$ and let k_1 be the length of the orbit of i_1 under σ. Let ψ_1 be the cycle $(\sigma(i_1), \ldots, \sigma^{k_1}(i_1))$. If $\psi_1 = \sigma$ the proof is complete. Otherwise there exists $i_2 \in N$, not in the orbit of i_1. Let $\psi_2 = (\sigma(i_2), \ldots, \sigma^{k_2}(i_2))$ where k_2 is the length of the orbit of i_2. Note that, by part (a), the cycles ψ_1 and ψ_2 are disjoint. Again, either $\sigma = \psi_2\psi_1$ and the proof is completed or there exists $i_3 \in N$, not in the union of orbits if i_1 and i_2, and so on. Since N is finite the process will terminate after a finite number of steps m. Then

$$\sigma = \psi_m \cdots \psi_1.$$

Observe that the elements permuted by a cycle ψ_t form an orbit under σ. The uniqueness of the decomposition of σ into disjoint cycles follows from part (a) and Defs. 4.3 and 4.4.

(c)   Clearly, by part (b), it suffices to prove that every cycle is a product of transpositions. The reader can easily check that

$$\sigma = (i_1, i_2, \ldots, i_k) = (i_1,i_k)(i_1,i_{k-1}) \cdots (i_1,i_2).$$

Definition 4.5   (*Even and odd permutations*)   *Let* $\sigma = \sigma_m \cdots \sigma_1$ $\in S_n$ *be the decomposition of* σ *into disjoint cycles. Let* σ_j *be a* k_j*-cycle,* $j = 1, \ldots, m$. *The number*

$$N(\sigma) = \sum_{j=1}^{m} (k_j - 1)$$

is called the Cauchy number of σ. *If every integer appears in exactly one of* $\sigma_1, \ldots, \sigma_m$, *then clearly*

$$N(\sigma) = n - m.$$

The permutation σ *is said to be* **even** *or* **odd** *according as its Cauchy number is even or odd. We define the* **signum** *of* σ *by*

$$\epsilon(\sigma) = (-1)^{N(\sigma)}.$$

Theorem 4.3   *If* $\sigma = \tau_k \tau_{k-1} \cdots \tau_1$, *where* $\tau_1, \ldots, \tau_k$ *are transpositions in* S_n, *then*

$$\epsilon(\sigma) = (-1)^k$$

i.e., σ *is even or odd according as* k *is.*

Proof.   We begin by showing that if a permutation σ is multiplied by a transposition τ then the signum changes. First, let σ be a cycle φ. Suppose that $\varphi = (i_1, \ldots, i_r, j_1, \ldots, j_s)$ and $\tau = (i_1, j_1)$. Then

$$\tau\varphi(i_t) = \tau(i_{t+1}) = i_{t+1}, \quad \text{for } t = 1, \ldots, r-1,$$
$$\tau\varphi(i_r) = \tau(j_1) = i_1,$$
$$\tau\varphi(j_t) = \tau(j_{t+1}) = j_{t+1}, \quad \text{for } t = 1, \ldots, s-1,$$

and

$$\tau\varphi(j_s) = \tau(i_1) = j_1.$$

Therefore

$$\tau\varphi = (i_1, \ldots, i_r)(j_1, \ldots, j_s)$$

and

$$N(\tau\varphi) = r + s - 2 = N(\varphi) - 1.$$

Now, suppose that only one of the numbers in the transposition is involved in the cycle. Let $\varphi = (i_1, \ldots, i_r)$ and $\tau = (i_1, j_1)$. Then

$$\tau\varphi = (i_1, \ldots, i_r, j_1)$$

and

$$N(\tau\varphi) = r = N(\varphi) + 1.$$

Lastly, if τ and φ are disjoint then, by the definition of the Cauchy number,

$$N(\tau\varphi) = N(\varphi) + N(\tau) = N(\varphi) + 1.$$

Suppose now that σ is a product of disjoint cycles, $\sigma = \varphi_1\varphi_2 \cdots \varphi_m$. If $\tau = (i_1,j_1)$ and i_1,j_1 appear in at most one of the φ_t then the result follows as above. If $\tau = (i_1,j_1)$, $\varphi_1 = (i_1, \ldots, i_r)$ and $\varphi_2 = (j_1, \ldots, j_s)$ then, as before, $\tau\varphi_1\varphi_2 = (i_1, \ldots, i_r, j_1, \ldots, j_s)$ so that

$$
\begin{aligned}
N(\tau\sigma) &= N(\tau\varphi_1\varphi_2) + N(\varphi_3 \cdots \varphi_m) \\
&= N(\varphi_1\varphi_2) + 1 + N(\varphi_3 \cdots \varphi_m) = N(\sigma) + 1.
\end{aligned}
$$

Now let $\sigma = \tau_k\sigma_1$, where $\sigma_1 = \tau_{k-1} \cdots \tau_1$, and use induction on k. If $k = 1$, σ is a transposition and $N(\sigma) = 1$. Otherwise

$$N(\sigma) = N(\sigma_1) \pm 1$$

But, by the induction hypothesis,

$$(-1)^{N(\sigma_1)} = (-1)^{k-1}.$$

Hence

$$\epsilon(\sigma) = (-1)^{N(\sigma)} = (-1)^k.$$

An immediate consequence of Theorem 4.3 is

Theorem 4.4   (a)   *If σ, $\varphi \in S_n$, then*

$$\epsilon(\sigma\varphi) = \epsilon(\sigma)\epsilon(\varphi)$$

and

$$\epsilon(\sigma^{-1}) = \epsilon(\sigma).$$

Proof. If σ is a product of h transpositions and φ is a product of k transpositions, then $\sigma\varphi$ is a product of $h + k$ transpositions and therefore

$$\epsilon(\sigma\varphi) = (-1)^{h+k} = (-1)^h(-1)^k = \epsilon(\sigma)\epsilon(\varphi).$$

Also,

$$1 = \epsilon(\sigma\sigma^{-1}) = \epsilon(\sigma)\epsilon(\sigma^{-1})$$

and thus

$$\epsilon(\sigma^{-1}) = 1/\epsilon(\sigma) = \epsilon(\sigma).$$

We conclude this section with three special results required later in our study of determinants.

Theorem 4.5 *Let $\sigma \in S_n$ and let $\sigma(i) = i$, $i = 1, \ldots, k$. Define $\varphi \in S_{n-k}$ by $\varphi(t) = \sigma(t + k) - k$, $t = 1, \ldots, n - k$. Then*

$$\epsilon(\sigma) = \epsilon(\varphi).$$

Proof. Let

$$\sigma = (1)(2) \cdots (k)\tau_m \cdots \tau_1$$

where $\tau_1, \ldots, \tau_m$ are transpositions involving the numbers $k + 1$, $\ldots, n$. For each $\tau_s = (i_s, j_s)$ set

$$\tau'_s = (i_s - k, j_s - k) \in S_{n-k}$$

so that, for $i = k + 1, \ldots, n$,

$$\tau'_1(i - k) = \tau_1(i) - k$$
$$\tau'_2\tau'_1(i - k) = \tau'_2(\tau_1(i) - k) = \tau_2\tau_1(i) - k$$
$$\cdots \cdots \cdots$$
$$\tau'_m \cdots \tau'_1(i - k) = \tau_m \cdots \tau_1(i) - k = \sigma(i) - k,$$
$$i = k + 1, \ldots, n,$$

and therefore

$$\tau'_m \cdots \tau'_1 = \varphi.$$

Hence, by Theorem 4.3,

$$\epsilon(\varphi) = (-1)^m = \epsilon(\sigma).$$

Theorem 4.6 *Let $1 \leq \alpha_1 < \cdots < \alpha_k \leq n$ and $1 \leq \beta_1 < \cdots <$*

$\beta_{n-k} \leq n$ be complementary sets of integers in the set $\{1, \ldots, n\}$, i.e.,

$$\{\alpha_1, \ldots, \alpha_k\} \cup \{\beta_1, \ldots, \beta_{n-k}\} = \{1, \ldots, n\}.$$

Let σ be the permutation in S_n defined by

$$\sigma(i) = \begin{cases} \alpha_i, & i = 1, \ldots, k, \\ \beta_{i-k}, & i = k+1, \ldots, n. \end{cases}$$

Then

$$\epsilon(\sigma) = (-1)^{s(\alpha) + k(k+1)/2}$$

where

$$s(\alpha) = \sum_{t=1}^{k} \alpha_t.$$

Proof.   The idea behind the proof of this theorem runs as follows. The sequence $(1, \ldots, n)$ can be changed into $(\alpha_1, 1, \ldots, \alpha_1 - 1, \alpha_1 + 1, \ldots, n)$ by $\alpha_1 - 1$ transpositions of adjacent numbers. Now, in a similar manner, α_2 can be brought by $\alpha_2 - 2$ transpositions into the second position leaving α_1 in the first position and the relative position of the remaining $n - 2$ numbers unchanged; and so on. The total number of transpositions required to build up σ is

$$(\alpha_1 - 1) + (\alpha_2 - 2) + \cdots + (\alpha_k - k) = s(\alpha) - k(k+1)/2$$

and therefore

$$\epsilon(\sigma) = (-1)^{s(\alpha) - k(k+1)/2}.$$

A somewhat altered rigorous version of this argument now follows. We use induction on n. If $n = 1$ there is nothing to prove. Assume that $n > 1$ and that the theorem holds for any permutation in S_{n-1} of the type described in the statement of the theorem. There are two possibilities: either $\beta_{n-k} = n$ or $\alpha_k = n$. If $\beta_{n-k} = n$, so that

$$\sigma = \begin{pmatrix} 1 & \cdots & k & k+1 & \cdots & n-1 & n \\ \alpha_1 & \cdots & \alpha_k & \beta_1 & \cdots & \beta_{n-k-1} & n \end{pmatrix},$$

we define $\tau \in S_{n-1}$ by

$$\tau(i) = \sigma(i), \quad i = 1, \ldots, n-1.$$

Clearly τ and σ have the same Cauchy number and therefore $\epsilon(\tau) = \epsilon(\sigma)$.

Also, by the induction hypothesis,

$$\epsilon(\tau) = (-1)^{s(\alpha) + k(k+1)/2}.$$

Now suppose that $\alpha_k = n$. Then

$$\sigma = \begin{pmatrix} 1 & \cdots & k-1 & k & k+1 & \cdots & n \\ \alpha_1 & \cdots & \alpha_{k-1} & n & \beta_1 & \cdots & \beta_{n-k} \end{pmatrix}.$$

Let

$$\theta = \begin{pmatrix} 1 & \cdots & k-1 & k & k+1 & \cdots & n-1 & n \\ \alpha_1 & \cdots & \alpha_{k-1} & \beta_1 & \beta_2 & \cdots & \beta_{n-k} & n \end{pmatrix}.$$

Then, as before, we can apply the induction hypothesis to obtain

$$\epsilon(\theta) = (-1)^{\alpha_1 + \cdots + \alpha_{k-1} + (k-1)k/2}.$$

Moreover,

$$\theta^{-1}\sigma = \begin{pmatrix} \alpha_1 & \cdots & \alpha_{k-1} & \beta_1 & \beta_2 & \cdots & \beta_{n-k} & n \\ 1 & \cdots & k-1 & k & k+1 & \cdots & n-1 & n \end{pmatrix}$$
$$\begin{pmatrix} 1 & \cdots & k-1 & k & k+1 & \cdots & n \\ \alpha_1 & \cdots & \alpha_{k-1} & n & \beta_1 & \cdots & \beta_{n-k} \end{pmatrix}$$
$$= \begin{pmatrix} 1 & 2 & \cdots & k-1 & k & k+1 & k+2 & \cdots & n-1 & n \\ 1 & 2 & \cdots & k-1 & n & k & k+1 & \cdots & n-2 & n-1 \end{pmatrix}$$
$$= (n, n-1, n-2, \ldots, k+1, k).$$

Thus

$$\epsilon(\theta^{-1}\sigma) = (-1)^{n-k} = (-1)^{\alpha_k - k}(-1)^{2k} = (-1)^{\alpha_k + k}$$

since $\alpha_k = n$ and $(-1)^{2k} = 1$. It follows that

$$\begin{aligned} \epsilon(\sigma) &= \epsilon(\theta\theta^{-1}\sigma) \\ &= \epsilon(\theta)\epsilon(\theta^{-1}\sigma) \\ &= (-1)^{\alpha_1 + \cdots + \alpha_{k-1} + (k-1)k/2}(-1)^{\alpha_k + k} \\ &= (-1)^{\alpha_1 + \cdots \alpha_{k-1} + \alpha_k + k + (k-1)k/2} \\ &= (-1)^{s(\alpha) + k(k+1)/2}. \end{aligned}$$

Theorem 4.7 *Let* $1 \leq j_1 < \cdots < j_k \leq n$ *and* $\sigma \in S_k$. *Let* $\varphi \in S_n$ *be defined by*

$$\begin{aligned} \varphi(j_t) &= j_{\sigma(t)}, \quad t = 1, \ldots, k, \\ \varphi(i) &= i, \quad i \notin \{j_1, \ldots, j_k\}. \end{aligned}$$

Then

$$\epsilon(\varphi) = \epsilon(\sigma).$$

Proof. Let $\sigma = \sigma_m \cdots \sigma_1$ be a decomposition of σ into transpositions. Now, define transpositions $\varphi_s \in S_n$, $s = 1, \ldots, m$, by

$$\varphi_s(j_t) = j_{\sigma_s(t)}, \quad t = 1, \ldots, k,$$
$$\varphi_s(i) = i, \quad i \notin \{j_1, \ldots, j_k\}.$$

Then

$$\varphi_m \cdots \varphi_1(j_t) = j_{\sigma_m \ldots \sigma_1(t)} = j_{\sigma(t)}, \quad t = 1, \ldots, k,$$
$$\varphi_m \cdots \varphi_1(i) = i, \quad i \notin \{j_1, \ldots, j_k\}.$$

Thus

$$\varphi_m \cdots \varphi_1 = \varphi$$

and

$$\epsilon(\varphi) = (-1)^m = \epsilon(\sigma).$$

Quiz

Answer true or false:

1. If $\sigma_1, \ldots, \sigma_m$ are transpositions in S_n, then $N(\sigma_m \cdots \sigma_1) = n - m$.
2. If $\sigma_1, \ldots, \sigma_m$ are transpositions in S_n, then

$$\epsilon(\sigma_m \cdots \sigma_1) = (-1)^{n-m}.$$

3. A power of a k-cycle is a k-cycle.
4. If $i_1 \le j_1 < j_2 \le i_2$ and $(i_1 i_2)(j_1 j_2) \ne (j_1 j_2)(i_2 i_1)$ then $i_1 = j_1$ or $i_2 = j_2$, but not both.
5. Any two 1-cycles in S_n are equal.
6. The kth power of a k-cycle is the identity permutation.
7. $(1 \quad 2)(1 \quad 3) = (1 \quad 2 \quad 3)$.
8. $(1 \quad 2 \quad 3 \quad 4)^{-1} = (2 \quad 3 \quad 4 \quad 1)$.
9. If $\sigma = (1 \quad 2 \quad 3 \quad 4)$ and $\tau = (1 \quad 3)(2 \quad 4)$ then $\tau\sigma\tau^{-1} = \sigma$.
10. If $\sigma = (1 \quad 2 \quad 3 \quad 4)$ and $\varphi = (1 \quad 4)(2 \quad 3)$ then $\varphi\sigma\varphi^{-1} = \sigma$.

Exercises

1. Prove that every even permutation in S_n, $n > 2$, is a product of 3-cycles.
2. Write the permutation

$$\sigma = \begin{pmatrix} 1 & 2 & 3 & 4 & 5 & 6 & 7 & 8 \\ 3 & 7 & 2 & 6 & 5 & 4 & 8 & 1 \end{pmatrix}$$

  (a) as a product of disjoint cycles,
  (b) as a product of transpositions.
3. Find the signum of the permutation σ in Exercise 2, using part (a) and check your answer by using part (b).
4. Let $\sigma \in S_n$ be a product of disjoint transpositions $\sigma = \sigma_m \cdots \sigma_1$. Then, by Def. 4.5, $\epsilon(\sigma) = (-1)^{N(\sigma)} = (-1)^{n-m}$. Explain this apparent discrepancy, since, by Theorem 4.3, $\epsilon(\sigma) = (-1)^m$.
5. Show that if $\sigma = (i_1 \cdots i_k)$ is a k-cycle in S_n and $\tau \in S_n$ then

$$\tau\sigma\tau^{-1} = (\tau(i_1) \cdots \tau(i_k)).$$

6. Show the number of even permutations in S_n is $n!/2$.
7. How many distinct k-cycles are there in S_n?

2.5 Introduction to Determinants

In this section we shall develop the properties of certain important scalar valued functions of matrices. These functions are of great theoretical and practical importance and appear throughout mathematics.

Definition 5.1 (Diagonal) *Let R be a field and $A \in M_n(R)$. If $\sigma \in S_n$ then the **diagonal** corresponding to σ is the n-tuple of numbers $(a_{1\sigma(1)}, a_{2\sigma(2)}, \ldots, a_{n\sigma(n)})$. A **diagonal product** corresponding to σ is just the product*

$$\prod_{i=1}^{n} a_{i\sigma(i)} = a_{1\sigma(1)}a_{2\sigma(2)} \cdots a_{n\sigma(n)}. \tag{1}$$

The matrix functions in this section will be of the following type: let χ be a function on S_n with values in R, i.e., a function that associates a scalar with each permutation in S_n. Then for $A \in M_n(R)$ we define $d_\chi(A)$ by

$$d_\chi(A) = \sum_{\sigma \in S_n} \chi(\sigma) \prod_{i=1}^{n} a_{i\sigma(i)}. \tag{2}$$

Definition 5.2   (**Permanent, Determinant**)   *The function* d_χ *that is obtained by setting* $\chi(\sigma) = 1$, $\sigma \in S_n$, *in* (2) *is called the* **permanent** *of A and is denoted by* per (A). *Thus*

$$\text{per } (A) = \sum_{\sigma \in S_n} \prod_{i=1}^{n} a_{i\sigma(i)}. \tag{3}$$

The function d_χ *obtained by setting* $\chi(\sigma) = \epsilon(\sigma)$, $\sigma \in S_n$, *in* (2) *is called the* **determinant** *of A and is denoted by* det (A). *Thus*

$$\det (A) = \sum_{\sigma \in S_n} \epsilon(\sigma) \prod_{i=1}^{n} a_{i\sigma(i)}. \tag{4}$$

Before we go into the special properties of these functions we want to consider them in a somewhat different way. Thus let $v_i = (a_{i1}, \ldots, a_{in})$, $i = 1, \ldots, n$, be a set of n vectors in $V_n(R)$. Let A be the matrix in $M_n(R)$ whose ith row is v_i, $i = 1, \ldots, n$. Then define a function on the set of n vectors $v_1, \ldots, v_n$ by

$$d_\chi(v_1, \ldots, v_n) = d_\chi(A). \tag{5}$$

As examples of (5) we can write

$$\det (A_{(1)}, \ldots, A_{(n)}) = \det (A), \text{ per } (A_{(1)}, \ldots, A_{(n)}) = \text{per } (A).$$

Theorem 5.1   *The function* d_χ *is linear in each vector variable separately. That is,*

$$
\begin{aligned}
d_\chi(v_1, \ldots, &v_{i-1}, r_1 v_i + r_2 v_i', v_{i+1}, \ldots, v_n) \\
&= r_1 d_\chi(v_1, \ldots, v_{i-1}, v_i, v_{i+1}, \ldots, v_n) \\
&\quad + r_2 d_\chi(v_1, \ldots, v_{i-1}, v_i', v_{i+1}, \ldots, v_n)
\end{aligned}
$$

for all vectors $v_1, \ldots, v_i, v_i', v_{i+1}, \ldots, v_n$ *in* $V_n(R)$ *and all scalars* r_1, r_2 *in* R.

Proof.   Let $v_k = (a_{k1}, \ldots, a_{kn})$, $k = 1, \ldots, n$, and set $v_i' = (a_{i1}', \ldots, a_{in}')$. In what follows, the notation $\prod_{t \neq i}$ means the product of the $n - 1$ indicated numbers obtained by setting $t = 1, \ldots, n$, with the exception of $t = i$. Then

$$d_\chi(v_1, \ldots, v_{i-1}, r_1 v_i + r_2 v_i', v_{i+1}, \ldots, v_n)$$

$$= \sum_{\sigma \in S_n} \chi(\sigma) \left(\prod_{t \neq i} a_{t\sigma(t)} \right) (r_1 a_{i\sigma(i)} + r_2 a_{i\sigma(i)}')$$

$$= r_1 \sum_{\sigma \in S_n} \chi(\sigma) \prod_{t=1}^{n} a_{t\sigma(t)} + r_2 \sum_{\sigma \in S_n} \chi(\sigma) \left(\prod_{t \neq i} a_{t\sigma(t)} \right) a_{i\sigma(i)}'$$

$$= r_1 d_\chi(v_1, \ldots, v_{i-1}, v_i, v_{i+1}, \ldots, v_n)$$
$$+ r_2 d_\chi(v_1, \ldots, v_{i-1}, v_i', v_{i+1}, \ldots, v_n).$$

As an immediate consequence of Theorem 5.1 we can write

$$\det \left(\sum_{t=1}^{k} c_{1t} u_t, \sum_{t=1}^{k} c_{2t} u_t, \ldots, \sum_{t=1}^{k} c_{nt} u_t \right)$$

$$= \sum_{t_1, \ldots, t_n = 1}^{k} c_{1t_1} c_{2t_2} \cdots c_{nt_n} \det (u_{t_1}, \ldots, u_{t_n}) \quad (6)$$

where $u_t \in V_n(R)$ and $c_{it} \in R$, $t = 1, \ldots, k$, $i = 1, \ldots, n$. In formula (6) the summation indices $t_1, \ldots, t_n$ each run independently from 1 to k, so that formally the sum involves k^n terms (many, of course, will be zero).

Another important general property of the d_χ function is contained in

Theorem 5.2   *If the function χ has the property that*

$$\chi(\sigma\varphi) = \chi(\sigma)\chi(\varphi) \quad (7)$$

for all σ and φ in S_n, then

$$d_\chi(v_{\varphi(1)}, \ldots, v_{\varphi(n)}) = \chi(\varphi) d_\chi(v_1, \ldots, v_n) \quad (8)$$

for any $\varphi \in S_n$.

Proof.   Once again we set $v_i = (a_{i1}, \ldots, a_{in})$, $i = 1, \ldots, n$, so that

$$d_\chi(v_{\varphi(1)}, \ldots, v_{\varphi(n)}) = \sum_{\sigma \in S_n} \chi(\sigma) \prod_{i=1}^{n} a_{\varphi(i),\sigma(i)}. \quad (9)$$

If we set $t = \varphi(i)$ in

$$\prod_{i=1}^{n} a_{\varphi(i),\sigma(i)}$$

then this term becomes

$$\prod_{t=1}^{n} a_{t,\sigma\varphi^{-1}(t)}.$$

Thus from (9)

$$d_{\chi}(v_{\varphi(1)}, \ldots, v_{\varphi(n)}) = \sum_{\sigma \in S_n} \chi(\sigma) \prod_{t=1}^{n} a_{t,\sigma\varphi^{-1}(t)}. \qquad (10)$$

We know from Theorem 4.1 (f), that as σ runs through S_n so does $\sigma\varphi^{-1}$ where φ is a fixed permutation in S_n. Thus in (10) we can set $\theta = \sigma\varphi^{-1}$, i.e., $\sigma = \theta\varphi$, and sum over θ instead of σ. We have

$$d_{\chi}(v_{\varphi(1)}, \ldots, v_{\varphi(n)}) = \sum_{\theta \in S_n} \chi(\theta\varphi) \prod_{t=1}^{n} a_{t,\theta(t)}. \qquad (11)$$

But, by (7), $\chi(\theta\varphi) = \chi(\theta)\chi(\varphi)$ so that (8) follows from (11).

We can conclude from Theorem 5.2 that

$$\text{per } (E_{(i),(j)}A) = \text{per } (A), \quad \det (E_{(i),(j)}A) = -\det (A). \qquad (12)$$

For, let φ be the transposition $(i \quad j)$. Then, since the value of $\chi(\varphi)$ is 1 for the permanent and -1 for the determinant, (12) follows immediately from (8). In fact, if $\varphi \in S_n$ then

$$\det (A_{(\varphi(1))}, \ldots, A_{(\varphi(n))}) = \epsilon(\varphi) \det (A_{(1)}, \ldots, A_{(n)}). \qquad (13)$$

We can also conclude from (6) with $\chi(\sigma) = \epsilon(\sigma)$ that

$$
\begin{aligned}
\det (E_{(i+c(j)}A) = \det (A_{(1)}, &\ldots, A_{(i-1)}, A_{(i)} + cA_{(j)}, \ldots, \\
&A_{(j)}, \ldots, A_{(n)}) \\
= \det (A_{(1)}, &\ldots, A_{(i-1)}, A_{(i)}, \ldots, A_{(n)}) \\
+ c \det (A_{(1)}, &\ldots, A_{(i-1)}, A_{(j)}, A_{(i+1)}, \ldots, \\
&A_{(j-1)}, A_{(j)}, \ldots, A_{(n)}) \\
= \det (A) + c \det (A_{(1)}, &\ldots, A_{(j)}, A_{(i+1)}, \ldots, \\
&A_{(j-1)}, A_{(j)}, \ldots, A_{(n)}).
\end{aligned}
$$

The last term on the right must be zero because when we interchange the two equal vectors $A_{(j)}$ appearing in the ith and jth positions the sign of the determinant function must change. On the other hand, the function obviously stays the same since the vectors are unaltered by this interchange. Thus it must be zero. We then have

$$\det (E_{(i) + c(j)}A) = \det (A). \tag{14}$$

In the above calculation we have assumed $j > i$. If $i > j$ the proof of (14) is essentially the same.

By precisely the same kind of calculation as before, the reader may verify that

$$\det (E_{c(i)}A) = c \det (A). \tag{15}$$

If we take $A = I_n$ in (12), (14), and (15) and observe that $\det (I_n) = 1$ we have

$$\det (E_{(i),(j)}) = -1, \quad \det (E_{(i) + c(j)}) = 1, \quad \det (E_{c(i)}) = c. \tag{16}$$

Moreover, if E is any of the three types of elementary matrices then it follows from looking at (12), (14), (15) and (16) that

$$\det (EA) = \det (E) \det (A). \tag{17}$$

We are now able to prove an important and classical result.

Theorem 5.3   *If R is a field and $A \in M_n(R)$, then A is singular if and only if* $\det (A) = 0$.

Proof.   Let $H \in M_n(R)$ be the Hermite normal form for A. If A is singular then the last row of H is zero (see Theorems 3.1 and 3.2). It follows that every diagonal product in ^tH is zero and hence $\det (H) = 0$. But $H = E_m E_{m-1} \cdots E_1 A$, where each E_i is an elementary matrix. Using (17) repeatedly we have

$$\det (H) = \det (E_m) \cdots \det (E_1) \det (A). \tag{18}$$

But, according to (16), $\det (E_j) \neq 0, j = 1, \ldots, m$, and hence (18) must imply that $\det (A) = 0$. Conversely, we see from (18) that if $\det (A) = 0$ then $\det (H) = 0$. But if A were nonsingular we would know from Theorem 3.2 (a) that $H = I_n$ and hence det $(H) =$ det $(I_n) = 1$. It follows that A is singular.

As an example of the use of Theorem 5.3 let us consider the homogeneous system

$$
\begin{aligned}
x_1 + 2x_2 + x_3 &= 0 \\
x_2 - 3x_3 &= 0 \\
-x_1 + x_2 - x_3 &= 0.
\end{aligned}
\tag{19}
$$

This becomes in matrix notation

$$Ax = 0$$

where

$$
A = \begin{bmatrix} 1 & 2 & 1 \\ 0 & 1 & -3 \\ -1 & 1 & -1 \end{bmatrix}.
$$

Now det $(A) = (-1 + 6 + 0) - (-1 - 3 + 0) = 9$. Hence A is nonsingular and, by Theorem 3.3, the system (19) has no solutions other than $x = 0$.

There is an interesting and simple fact that can be quickly verified for the permanent and determinant. Namely, if $A \in M_n(R)$, then

$$
\det (A^T) = \det (A), \quad \operatorname{per} (A^T) = \operatorname{per} (A).
\tag{20}
$$

To verify (20), we note first that

$$
\begin{aligned}
\det (A^T) &= \sum_{\sigma \in S_n} \epsilon(\sigma) \prod_{i=1}^{n} (A^T)_{i\sigma(i)} \\
&= \sum_{\sigma \in S_n} \epsilon(\sigma) \prod_{i=1}^{n} a_{\sigma(i)i}.
\end{aligned}
$$

In $\prod_{i=1}^{n} a_{\sigma(i)i}$ the order of the factors is immaterial. Thus set $j = \sigma(i)$ so that $i = \sigma^{-1}(j)$ and then

$$
\prod_{i=1}^{n} a_{\sigma(i)i} = \prod_{j=1}^{n} a_{j\sigma^{-1}(j)}.
$$

Hence

$$\det (A^T) = \sum_{\sigma \in S_n} \epsilon(\sigma) \prod_{j=1}^{n} a_{j\sigma^{-1}(j)}.$$

Set $\varphi = \sigma^{-1}$. Since φ runs over S_n as σ does and $\epsilon(\varphi) = \epsilon(\sigma^{-1}) = \epsilon(\sigma)$, we have

$$\det (A^T) = \sum_{\varphi \in S_n} \epsilon(\varphi) \prod_{j=1}^{n} a_{j\varphi(j)} = \det (A).$$

The reader is asked to show similarly that per (A^T) = per (A) in Exercise 4.

The function per (A) has interesting combinatorial (i.e., enumerative) applications. For example, suppose we wish to count the number (call it p) of permutations σ in S_3 that satisfy $\sigma(i) \neq i$, $i = 1, 2, 3$. In other words, σ does not hold any integer fixed. Of course, this problem can be done easily by a direct count. However, the following method works for a general n, $n \geq 3$. Consider the matrix

$$A = \begin{bmatrix} 0 & 1 & 1 \\ 1 & 0 & 1 \\ 1 & 1 & 0 \end{bmatrix}.$$

Each $\sigma \in S_3$ corresponds to a term in the permanent expansion of A, i.e., the diagonal product corresponding to σ. Moreover, a diagonal product has value 1 if and only if $\sigma(i) \neq i$, $i = 1, 2, 3$, i.e., the diagonal does not "hit" the zeros. Hence $p = $ per $(A) = 2$. For $n \geq 3$, we can define A to be the matrix with 0 on the main diagonal and 1 elsewhere. We state without proof that

$$\text{per } (A) = n! \sum_{j=0}^{n} \frac{(-1)^j}{j!}.$$

In order to handle effectively the usual hideous array of indices that creep into this subject we are going to introduce some notation for sequence sets.

Definition 5.3   (**Sequence sets**)   *Let r and n be positive integers. Define $\Gamma_{r,n}$ to be the totality of sequences α, $\alpha = (\alpha_1, \ldots, \alpha_r)$, in which* *each α_i is an integer satisfying $1 \leq \alpha_i \leq n$. Next let $G_{r,n}$ be the subset of $\Gamma_{r,n}$ consisting of precisely those sequences α for which $1 \leq \alpha_1 \leq \alpha_2 \leq \cdots \leq \alpha_r \leq n$. Finally, if $r \leq n$ then $Q_{r,n}$ will denote the subset of $G_{r,n}$ consisting of those α for which $1 \leq \alpha_1 < \cdots < \alpha_r \leq n$. Thus $\Gamma_{r,n}$ is the set of all sequences of length r chosen from $1, \ldots, n$; $G_{r,n}$ is the set of non-decreasing sequences in $\Gamma_{r,n}$, and $Q_{r,n}$ is the set of strictly increasing sequences in $G_{r,n}$.*

Definition 5.4 **(Submatrix)** *Let R be a field and $A \in M_{m,n}(R)$. Let $\alpha \in \Gamma_{r,m}$ and $\beta \in \Gamma_{s,n}$. Then $A[\alpha|\beta]$ is the matrix in $M_{r,s}(R)$ whose (i,j) entry is $a_{\alpha_i \beta_j}$, $i = 1, \ldots, r$, $j = 1, \ldots, s$. If α and β are in $Q_{r,m}$ and $Q_{s,n}$ respectively then $A[\alpha|\beta]$ is called a **submatrix** of A. The determinant of a square submatrix is called a **subdeterminant**. For $\alpha \in Q_{r,m}$ and $\beta \in Q_{s,n}$ let α' and β' denote the sequences in $Q_{m-r,m}$ and and $Q_{n-s,n}$ whose integers are complementary to α and β respectively. Then we define $A(\alpha|\beta) = A[\alpha'|\beta']$, $A[\alpha|\beta) = A[\alpha|\beta']$, $A(\alpha|\beta] = A[\alpha'|\beta]$. If $m = n$, $r = s$ and $\alpha \in Q_{r,n}$ then $A[\alpha|\alpha]$ is called a **principal** submatrix of A. The determinant of a principal submatrix is called a **principal subdeterminant**.*

For example, let

$$A = \begin{bmatrix} a_{11} & a_{12} & a_{13} \\ a_{21} & a_{22} & a_{23} \\ a_{31} & a_{32} & a_{33} \end{bmatrix}$$

and $\alpha = (2,2,3) \in \Gamma_{3,3}$, $\beta = (1,1) \in \Gamma_{2,3}$. Then

$$A[\alpha|\beta] = A[2,2,3|1,1] = \begin{bmatrix} a_{21} & a_{21} \\ a_{21} & a_{21} \\ a_{31} & a_{31} \end{bmatrix}.$$

Again, if $\alpha = (1,2) \in Q_{2,3}$, $\beta = (2,3) \in Q_{2,3}$ then

$$A[\alpha|\beta] = \begin{bmatrix} a_{12} & a_{13} \\ a_{22} & a_{23} \end{bmatrix},$$

$$A(\alpha|\beta] = A[\alpha'|\beta] = A[3|2,3]$$
$$= [a_{32} \quad a_{33}],$$

and

$$A(\alpha|\beta) = A[\alpha'|\beta'] = A[3|1] = [a_{31}].$$

Quiz

Answer true or false (in what follows, R is always a field):

1. If $A \in M_n(R)$ and every diagonal product is 0, then A has a zero row.
2. If $A \in M_n(R)$ then per $(A^2) = [$per $(A)]^2$ for any A.
3. If two rows of A are the same then per $(A) = 0$.
4. If $e_1, \ldots, e_n$ are the standard vectors in $V_n(R)$ then det $(e_{\varphi(1)}, \ldots, e_{\varphi(n)}) = \epsilon(\varphi)$ for any $\varphi \in S_n$.
5. There are no matrices for which det $(A) =$ per $(A) \neq 0$.
6. If $1 \le r \le n$ and $\alpha \in Q_{r,n}$ then $\alpha' \in Q_{n-r,n}$.
7. If $A = (a_{ij}) \in M_4(R)$ then

$$A[1,1|3,4] = \begin{bmatrix} a_{13} & a_{14} \\ a_{41} & a_{31} \end{bmatrix}.$$

8. If $A = (a_{ij}) \in M_{2,3}(R)$ then $A(1|2,3) = [a_{21}]$.
9. If $A \in M_7(R)$ and $\alpha = (1,2,3,4)$ then

$$A[\alpha|\alpha] = \begin{bmatrix} a_{15} & a_{16} & a_{17} \\ a_{25} & a_{26} & a_{27} \\ a_{35} & a_{36} & a_{37} \\ a_{45} & a_{46} & a_{47} \end{bmatrix}.$$

10. If $A \in M_n(R)$, $1 \le r < n$, $\alpha \in Q_{r,n}$ then $A[\alpha'|\alpha'] = A(\alpha|\alpha)$.

Exercises

1. Compute per (J_n) and det (J_n) where each entry of J_n is n^{-1} and J_n is $n \times n$.
2. Show that if two rows or columns of $A \in M_n(R)$ are the same then det $(A) = 0$.
3. Show that the system of equations

$$x_1 - x_2 = 0$$
$$x_2 + x_3 = 0$$
$$x_3 + x_4 = 0$$
$$x_4 + x_1 = 0$$

has only the solution $x_1 = x_2 = x_3 = x_4 = 0$.
4. Show that per $(A^T) =$ per (A) for $A \in M_n(R)$.
5. Find the number of permutations $\sigma \in S_4$ for which $\sigma(i) \neq i, i = 1,2,3,4$.
6. Show that if A is an n-square complex matrix and $b_{ij} = |a_{ij}|$, $i, j = 1, \ldots, n$, then

$$|\text{per } (A)| \le \text{per } (B).$$

7. Show that the number of sequences in $\Gamma_{r,n}$ is n^r.

8. Show that if $1 \leq r \leq n$ then the number of sequences in $Q_{r,n}$ is $\binom{n}{r}$.

9. Show that if $1 \leq r \leq n$ then the number of sequences in $G_{r,n}$ is

$$\binom{n+r-1}{r}.$$

10. How many $r \times s$ submatrices of an $m \times n$ matrix are there, $1 \leq r \leq m$, $1 \leq s \leq n$?

2.6 Properties of Determinants

In this section we shall obtain two major classical results about determinants and exploit some of their consequences.

The first result, known as the *Cauchy-Binet Theorem*, shows how to calculate the determinant of a submatrix (i.e., a subdeterminant) of a product AB if we know the subdeterminants of both factors A and B. All matrices in this section are over a fixed field R.

Theorem 6.1 *Let $A \in M_{p,q}(R)$ and $B \in M_{q,r}(R)$ and $1 \leq m \leq$ min (p,q,r). If $\alpha \in Q_{m,p}$, $\beta \in Q_{m,r}$, then*

$$\det\{(AB)[\alpha|\beta]\} = \sum_{\omega \in Q_{m,q}} \det\{A[\alpha|\omega]\} \det\{B[\omega|\beta]\}. \qquad (1)$$

In particular if $p = q = r = m$, it follows that

$$\det(AB) = \det(A)\det(B). \qquad (2)$$

Proof. Let $C = (AB)[\alpha|\beta]$. The (i,j) entry of C is the (α_i, β_j) entry of AB, i.e., it is the product of row α_i of A and column β_j of B:

$$c_{ij} = A_{(\alpha_i)}B^{(\beta_j)}, \quad i,j = 1, \ldots, m.$$

Thus the ith row of C is just

$$\begin{aligned}
C_{(i)} &= (A_{(\alpha_i)}B^{(\beta_1)}, A_{(\alpha_i)}B^{(\beta_2)}, \ldots, A_{(\alpha_i)}B^{(\beta_m)}) \\
&= \left(\sum_{t=1}^{q} a_{\alpha_i t}b_{t\beta_1}, \sum_{t=1}^{q} a_{\alpha_i t}b_{t\beta_2}, \ldots, \sum_{t=1}^{q} a_{\alpha_i t}b_{t\beta_m} \right) \\
&= \sum_{t=1}^{q} a_{\alpha_i t}(b_{t\beta_1}, \ldots, b_{t\beta_m}).
\end{aligned}$$

If we let $D = B[1, \ldots, q|\beta]$, then the preceding calculation shows that

$$C_{(i)} = \sum_{t=1}^{q} a_{\alpha_i t} D_{(t)}.$$

Hence, using formula (6) of Sec. 5, we have

$$\det (C) = \det \Big(\sum_{t=1}^{q} a_{\alpha_1 t} D_{(t)}, \ldots, \sum_{t=1}^{q} a_{\alpha_m t} D_{(t)} \Big) \tag{3}$$

$$= \sum_{(t_1, \ldots, t_m) \in \Gamma_{m,q}} \prod_{j=1}^{m} a_{\alpha_j t_j} \det (D_{(t_1)}, \ldots, D_{(t_m)}).$$

The terms that appear in the sum in (3) arising from choices of $(t_1, \ldots, t_m)$, in which $t_i = t_j$, for some $i \neq j$, drop out because the determinant of a matrix with two rows the same is zero. We can obtain all sequences of distinct integers $t_1, \ldots, t_m$ by choosing first an increasing sequence, i.e., $(t_1, \ldots, t_m) \in Q_{m,q}$, and then summing over all rearrangements of each such sequence. In other words, we have, by equation (13) in Sec. 5,

$$\det (C) = \sum_{(t_1, \ldots, t_m) \in Q_{m,q}} \sum_{\sigma \in S_m} \prod_{j=1}^{m} a_{\alpha_j t_{\sigma(j)}} \det (D_{(t_{\sigma(1)})}, \ldots, D_{(t_{\sigma(m)})})$$

$$\tag{4}$$

$$= \sum_{(t_1, \ldots, t_m) \in Q_{m,q}} \det (D_{(t_1)}, \ldots, D_{(t_m)}) \sum_{\sigma \in S_m} \epsilon(\sigma) \prod_{j=1}^{m} a_{\alpha_j t_{\sigma(j)}}.$$

Recall that $D = B[1, \ldots, q|\beta]$ and hence $\det (D_{(t_1)}, \ldots, D_{(t_m)}) = \det (B[t_1, \ldots, t_m|\beta])$.

By the definition of a determinant we have

$$\sum_{\sigma \in S_m} \epsilon(\sigma) \prod_{j=1}^{m} a_{\alpha_j t_{\sigma(j)}} = \det (A[\alpha|t_1, \ldots, t_m]).$$

Hence, upon setting $\omega = (t_1, \ldots, t_m) \in Q_{m,q}$ in (4) we have

$$\det \{(AB)[\alpha|\beta]\} = \det (C) = \sum_{\omega \in Q_{m,q}} \det (A[\alpha|\omega]) \det (B[\omega|\beta]).$$

If $p = q = r = m$ then there are only the choices $(1, \ldots, m) = \alpha = \beta = \omega$ in the formula (1) and, of course, we immediately obtain (2). We can illustrate Theorem 6.1 as follows. Let

$$A = \begin{bmatrix} 1 & 0 & 3 & 4 \\ 2 & 0 & -1 & 2 \end{bmatrix} \quad \text{and} \quad B = \begin{bmatrix} 1 & 1 \\ 1 & -1 \\ 2 & 2 \\ 0 & 0 \end{bmatrix}.$$

Then AB is a 2×2 matrix and we can compute its determinant by

$$\det (AB) = \sum_{\omega \in Q_{2,4}} \det (A[1,2|\omega]) \det (B[\omega|1,2]).$$

Now

$$\det (A[1,2|\omega]) \det (B[\omega|1,2]) = 0$$

whenever ω involves either a 2 or a 4, i.e., $A[1,2|\omega]$ has a zero column or $B[\omega|1,2]$ has a zero row. The only term that survives is the one obtained by setting $\omega = (1,3)$. Hence

$$\begin{aligned} \det (AB) &= \det (A[1,2|1,3]) \det (B[1,3|1,2]) \\ &= (-7)(0) = 0. \end{aligned}$$

Our next result is preliminary to Theorem 6.3, the so-called *Laplace Expansion Theorem*. It is in fact a special case of Theorem 6.3.

Theorem 6.2 *Let* $A \in M_n(R)$ *be a matrix of the following form:*

$$A = I_k \dotplus B \tag{5}$$

where $B \in M_{n-k}(R)$. *Then* $\det (A) = \det (B)$.

Proof: By definition,

$$\det (A) = \sum_{\sigma \in S_n} \epsilon(\sigma) \prod_{i=1}^{n} a_{i\sigma(i)}.$$

Clearly, unless $\sigma(i) = i, i = 1, \ldots, k$, the product $\displaystyle\prod_{i=1}^{n} a_{i\sigma(i)}$ is zero.

Thus

$$\det (A) = \sum_{\sigma \in S_n{}'} \epsilon(\sigma) \prod_{i=1}^{n} a_{i\sigma(i)}$$

where the dash over S_n indicates that the summation is taken only over those σ for which $\sigma(i) = i$, $i = 1, \ldots, k$. But since $a_{i\sigma(i)} = 1$, $i = 1, \ldots, k$, we have

$$\det (A) = \sum_{\sigma \in S_n{}'} \epsilon(\sigma) \prod_{i=k+1}^{n} a_{i\sigma(i)}.$$

Now $a_{k+s,k+t} = b_{st}$, $s, t = 1, \ldots, n - k$. Moreover for each $\sigma \in S_n'$ we can associate a unique $\varphi \in S_{n-k}$ defined by $\varphi(t) = \sigma(k + t) - k$ and satisfying $\epsilon(\varphi) = \epsilon(\sigma)$ (see Theorem 4.5).

Then

$$b_{s\varphi(s)} = a_{k+s,k+\varphi(s)} = a_{k+s,\sigma(k+s)},$$

and

$$\epsilon(\sigma) \prod_{i=k+1}^{n} a_{i\sigma(i)} = \epsilon(\sigma) \prod_{s=1}^{n-k} a_{k+s,\sigma(k+s)}$$

$$= \epsilon(\varphi) \prod_{s=1}^{n-k} b_{s\varphi(s)}.$$

Hence

$$\det (A) = \sum_{\varphi \in S_{n-k}} \epsilon(\varphi) \prod_{s=1}^{n-k} b_{s\varphi(s)}$$

$$= \det (B).$$

We are now in a position to prove a very important theorem that allows us to compute the determinant of a matrix in terms of determinants of smaller submatrices. This result is called the *Laplace Expansion Theorem*.

Theorem 6.3 *Let $A \in M_n(R)$, and let α be a fixed sequence in $Q_{k,n}$, $1 \leq k \leq n$.   Then*

$$\det (A) = \sum_{\gamma \in Q_{k,n}} (-1)^{s(\alpha) + s(\gamma)} \det (A[\alpha|\gamma]) \det (A(\alpha|\gamma)) \qquad (6)$$

where

$$s(\gamma) = \sum_{j=1}^{k} \gamma_j$$

for any $\gamma \in Q_{k,n}$.

Proof.   We write $\det (A) = \det (A_{(1)}, \ldots, A_{(n)})$, and then set

$$A_{(t)} = \sum_{j=1}^{n} a_{tj} e_j, \quad t = 1, \ldots, n,$$

where as usual the e_j are the standard vectors with 1 in position j, zero elsewhere. Replacing $A_{(\alpha_j)}$ by

$$\sum_{t=1}^{n} a_{\alpha_j t} e_t, \quad j = 1, \ldots, k,$$

we have

$$\det (A) = \det (A_{(1)}, \ldots, A_{(\alpha_1)}, \ldots, A_{(\alpha_k)}, \ldots, A_{(n)}) \qquad (7)$$

$$= \det (A_{(1)}, \ldots, \sum_{j=1}^{n} a_{\alpha_1 j} e_j, \ldots, \sum_{j=1}^{n} a_{\alpha_k j} e_j, \ldots, A_{(n)})$$

$$= \sum_{(j_1, \ldots, j_k) \in \Gamma_{k,n}} a_{\alpha_1 j_1} \cdots a_{\alpha_k j_k} \det (A_{(1)}, \ldots, e_{j_1}, \ldots,$$

$$e_{j_k}, \ldots, A_{(n)}).$$

In (7) we have only used rows $\alpha_1, \ldots, \alpha_k$ to expand. All others are left alone. In the last summation in (7) if any two of $j_1, \ldots, j_k$ are equal then of course $\det (A_{(1)}, \ldots, e_{j_1}, \ldots, e_{j_k}, \ldots, A_{(n)}) = 0$. Hence we need only sum over sets of distinct $j_1, \ldots, j_k$. Moreover, we can do this by choosing an increasing sequence $j_1 < \cdots < j_k$ and then summing over all $k!$ rearrangements of this sequence:

$$(j_{\sigma(1)}, \ldots, j_{\sigma(k)}), \quad \sigma \in S_k.$$

From Theorem 4.7 we know that

$$\epsilon \begin{pmatrix} j_1 & \cdots & j_k \\ j_{\sigma(1)} & \cdots & j_{\sigma(k)} \end{pmatrix} = \epsilon(\sigma).$$

Moreover, by the statement (13) of Sec. 5, it follows that

$$\det (A_{(1)}, \ldots, e_{j_{\sigma(1)}}, \ldots, e_{j_{\sigma(k)}}, \ldots, A_{(n)})$$
$$= \epsilon(\sigma) \det (A_{(1)}, \ldots, e_{j_1}, \ldots, e_{j_k}, \ldots, A_{(n)}). \quad (8)$$

Using (7) and (8), and setting $\gamma = (j_1, \ldots, j_k) \in Q_{k,n}$ we can write

$$\det (A) = \sum_{\gamma \in Q_{k,n}} \sum_{\sigma \in S_k} a_{\alpha_1 j_{\sigma(1)}} \cdots a_{\alpha_k j_{\sigma(k)}} \det (A_{(1)}, \ldots,$$
$$e_{j_{\sigma(1)}}, \ldots, e_{j_{\sigma(k)}}, \ldots, A_{(n)}) \quad (9)$$
$$= \sum_{\gamma \in Q_{k,n}} \left(\sum_{\sigma \in S_k} \epsilon(\sigma) a_{\alpha_1 j_{\sigma(1)}} \cdots a_{\alpha_k j_{\sigma(k)}} \right) \det (A_{(1)}, \ldots,$$
$$e_{j_1}, \ldots, e_{j_k}, \ldots, A_{(n)})$$
$$= \sum_{\gamma \in Q_{k,n}} \det A[\alpha|j_1, \ldots, j_k] \det (A_{(1)}, \ldots,$$
$$e_{j_1}, \ldots, e_{j_k}, \ldots, A_{(n)}).$$

We must now examine the determinant

$$\det (A_{(1)}, \ldots, e_{j_1}, \ldots, e_{j_k}, \ldots, A_{(n)}), \quad (10)$$

appearing in the last summation in (9). Rows numbered $\alpha_1, \ldots, \alpha_k$ may be brought into positions $1, \ldots, k$ by a permutation of the rows whose sign is $(-1)^{s(\alpha) - k(k+1)/2}$ (see Theorem 4.6). But e_{j_t} appears in row numbered α_t of (10). Thus we must examine the value of

$$(-1)^{s(\alpha) - k(k+1)/2} \det (e_{j_1}, \ldots, e_{j_k}, A_{(\beta_1)}, \ldots, A_{(\beta_{n-k})}) \quad (11)$$

where the rows following the kth are rows $A_{(t)}$ of A in which t is in the set complementary to $\alpha_1, \ldots, \alpha_k$ in $1, \ldots, n$; we have called this set $\beta_1, \ldots, \beta_{n-k}$ in (11).

We write out the matrix in (11) in terms of its rows:

$$
\begin{bmatrix}
e_{j_1} \\
e_{j_2} \\
\cdot \\
\cdot \\
\cdot \\
e_{j_k} \\
A_{(\beta_1)} \\
\cdot \\
\cdot \\
\cdot \\
A_{(\beta_{n-k})}
\end{bmatrix}.
\tag{12}
$$

By performing elementary row operations of type II on the matrix in (12) it is clear that without altering the value of the determinant we can annihilate the entries appearing in columns $j_1, \ldots, j_k$ and in rows $k + 1, \ldots, n$. We can then move columns $j_1, \ldots, j_k$ into column positions $1, \ldots, k$ thereby introducing a total of

$$
\sum_{t=1}^{k} j_t - k(k+1)/2 = s(\gamma) - k(k+1)/2
$$

sign changes in the determinant. The resulting matrix looks like

$$
\left[
\begin{array}{c|c}
I_k & 0 \\
\hline
0 & A(\alpha|j_1, \ldots, j_k)
\end{array}
\right] = I_k \dotplus A(\alpha|\gamma),
\tag{13}
$$

which, by Theorem 6.2, has determinant equal to $\det A(\alpha|\gamma)$. Thus from (9) and (11) we have

$$
\det(A) = \sum_{\gamma \in Q_{k,n}} (-1)^{s(\alpha) - k(k+1)/2} \det(A[\alpha|\gamma])
$$
$$
(-1)^{s(\gamma) - k(k+1)/2} \det(A(\alpha|\gamma))
$$
$$
= \sum_{\gamma \in Q_{k,n}} (-1)^{s(\alpha) + s(\gamma)} \det(A[\alpha|\gamma]) \det(A(\alpha|\gamma)).
$$

This latter formula follows because $k(k + 1)$ is always even. The proof is over but because of its complexity the student should recapitulate it until he understands what is going on.

Observe that if $k = 1$ and α is a single integer, then (6) collapses to

$$\det (A) = \sum_{\beta = 1}^{n} (-1)^{\alpha + \beta} a_{\alpha\beta} \det (A(\alpha|\beta)). \tag{14}$$

This is called the *expansion by row* α.

Since $\det (A) = \det (A^T)$ we can apply the Laplace Expansion Theorem to A^T and obtain an analogous theorem on expansion by columns. To do this we observe that $(A^T[\alpha|\beta])_{ij} = (A^T)_{\alpha_i, \beta_j} = A_{\beta_j, \alpha_i}$. Now, the (i,j) entry of $(A[\beta|\alpha])^T$ is the (j,i) entry of $A[\beta|\alpha]$, namely A_{β_j, α_i}. Hence

$$A^T[\alpha|\beta] = (A[\beta|\alpha])^T; \quad \det (A^T[\alpha|\beta]) = \det (A[\beta|\alpha]). \tag{15}$$

Also, the (i,j) entry of $A^T(\alpha|\beta)$ is the α'_i, β'_j entry of A^T where α' is the ordered complementary set to α in $1, \ldots, n$. Thus $(A^T(\alpha|\beta))_{i,j} = (A^T)_{\alpha'_i, \beta'_j} = A_{\beta'_j, \alpha'_i}$. The (i,j) entry of $(A(\beta|\alpha))^T$ is the (j,i) entry of $A(\beta|\alpha)$, i.e., $A_{\beta'_j, \alpha'_i}$. Hence

$$A^T(\alpha|\beta) = (A(\beta|\alpha))^T; \quad \det (A^T(\alpha|\beta)) = \det (A(\beta|\alpha))$$

and we can apply the Laplace Expansion Theorem to A^T to obtain

$$\begin{aligned} \det (A) &= \det (A^T) \\ &= \sum_{\gamma \in Q_{k,n}}' (-1)^{s(\alpha) + s(\gamma)} \det (A^T[\alpha|\gamma]) \det (A^T(\alpha|\gamma)) \\ &= \sum_{\gamma \in Q_{k,n}}' (-1)^{s(\alpha) + s(\gamma)} \det (A[\gamma|\alpha]) \det (A(\gamma|\alpha)). \end{aligned} \tag{16}$$

An expansion by columns which parallels the expansion (14) by rows works in essentially the same way as the expansion by rows.

As an example of the use of (16) we compute $\det (A)$ where

$$A = \begin{bmatrix} 0 & 1 & 2 & 3 \\ 1 & 0 & -1 & 4 \\ 0 & 0 & 7 & 5 \\ -1 & 0 & 2 & 8 \end{bmatrix}.$$

It is appropriate to use the Laplace expansion on the first two columns. Thus let $\alpha = (1,2)$ so that $(-1)^{s(\alpha)} = -1$. Clearly if 3 appears in γ then $\det (A[\gamma|1,2]) = 0$. The only γ that survive are $(1,2)$, $(1,4)$,

(2,4). Thus (16) becomes

$$\det (A) = -[(-1)^3(-1)(46)$$
$$+ (-1)^5(1)(-33) + (-1)^6(0)(-11)] = -79.$$

Definition 6.1 (Adjugate) *Let* $A \in M_n(R)$. *Define a matrix* $B \in M_n(R)$ *by*

$$b_{ij} = (-1)^{i+j} \det (A(j|i)), \; i,j = 1, \ldots , n. \qquad (17)$$

*Then B is called the **adjugate** of A and is denoted by* $B = \operatorname{adj} A$.

We are sorry to report that adj A is also called the adjoint of A. We, however, shall not call it by this name since the term adjoint already has been used with an entirely different meaning for linear transformations.

Theorem 6.4 *If* $A \in M_n(R)$ *then*

$$A(\operatorname{adj} A) = (\operatorname{adj} A)A = \det (A)I_n. \qquad (18)$$

Hence A is nonsingular if and only if $\det (A) \neq 0$. *If A is nonsingular, then*

$$A^{-1} = [\det (A)]^{-1} \operatorname{adj} A. \qquad (19)$$

Proof. Let $B = \operatorname{adj} A$ so that the (i,j) entry of AB is

$$\sum_{t=1}^{n} a_{it}b_{tj} = \sum_{t=1}^{n} a_{it}(-1)^{t+j} \det (A(j|t)).$$

If $i = j$, then this last summation becomes

$$\sum_{t=1}^{n} (-1)^{t+i}a_{it} \det (A(i|t))$$

which by (14) is precisely the expansion of the determinant of A using row i. It remains to show that, for $j \neq i$, $(AB)_{ji} = 0$. Suppose then that $i \neq j$ and consider the matrix C which agrees with A except that the ith row of C is the jth row of A. It is clear that $A(i|t) = C(i|t)$, $t = 1$,

. . . , n. Also C has two rows the same (both row i and row j of C are $A_{(j)}$) and hence det $(C) = 0$. Thus expanding C by its ith row we have

$$
\begin{aligned}
0 = \det (C) &= \sum_{t=1}^{n} c_{it}(-1)^{i+t} \det (C(i|t)) \\
&= \sum_{t=1}^{n} a_{jt}(-1)^{i+t} \det (A(i|t)) \\
&= \sum_{t=1}^{n} a_{jt}b_{ti} \\
&= (AB)_{ji}.
\end{aligned}
$$

Hence $AB = \det (A)I_n$. A similar argument will show that $BA = \det (A)I_n$. If det $(A) \neq 0$ then, by Theorem 5.3, A has an inverse and since the inverse is unique

$$
A^{-1} = [\det (A)]^{-1} \operatorname{adj} A.
$$

For example, if $A = \begin{bmatrix} a & b \\ c & d \end{bmatrix}$ and $ad - bc \neq 0$, we can write down A^{-1} from (19):

$$
A^{-1} = \frac{1}{ad - bc} \begin{bmatrix} d & -b \\ -c & a \end{bmatrix}.
$$

Theorem 6.4 will directly yield formulas (known to every schoolboy as *Cramer's Rule*) for the solution of certain linear equations. These formulas are unfortunately quite useless for any kind of numerical calculation except in rather special cases.

Theorem 6.5   *Let $A \in M_n(R)$ be a nonsingular matrix and suppose that x and b are n-tuples. If*

$$
Ax = b, \quad \text{i.e.,} \quad \sum_{t=1}^{n} a_{it}x_t = b_i, \quad i = 1, \ldots, n,
$$

then

$$
x_t = [\det (A)]^{-1} \det (A^{(1)}, \ldots, A^{(t-1)}, b, A^{(t+1)}, \ldots, A^{(n)}),
$$
$$
t = 1, \ldots, n. \quad (20)
$$

Proof.   Since A is nonsingular, det $(A) \neq 0$ and $A^{-1} = (\det A)^{-1}$ adj A. Thus

$$x = A^{-1}b = [\det (A)]^{-1}(\text{adj } A)b.$$

The tth entry in x is given by

$$
\begin{aligned}
x_t &= [\det (A)]^{-1}(\text{adj } A)_{(t)}b \\
&= [\det (A)]^{-1} \sum_{k=1}^{n} (\text{adj } A)_{tk}b_k \\
&= [\det (A)]^{-1} \sum_{k=1}^{n} (-1)^{t+k} \det (A(k|t))b_k.
\end{aligned}
$$

The summation in the last expression is just the Laplace expansion by column t of

$$\det (A^{(1)}, \ldots , A^{(t-1)}, b, A^{(t+1)}, \ldots , A^{(n)})$$

and the result follows.

We observe here that Theorem 6.5 can be used to compute particular columns of A^{-1}. For, we know that if we write $Ax = e_j$ then $x = A^{-1}e_j = (A^{-1})^{(j)}$. But (20) tells us that

$$
\begin{aligned}
x_t &= [\det (A)]^{-1} \det (A^{(1)}, \ldots , A^{(t-1)}, e_j, A^{(t+1)}, \ldots , A^{(n)}) \\
&= (-1)^{j+t}[\det (A)]^{-1} \det (A(j|t)), \qquad t = 1, \ldots , n.
\end{aligned}
$$

In other words, the jth column of A^{-1} is given by

$$(A^{-1})^{(j)} = [\det (A)]^{-1}((-1)^{j+1} \det (A(j|1)), \ldots ,$$
$$(-1)^{j+n} \det (A(j|n))). \quad (21)$$

For example, if

$$A = \begin{bmatrix} 1 & 1 & 0 \\ 0 & 1 & -1 \\ 1 & 0 & 3 \end{bmatrix}$$

then the first column of A^{-1} is

$$[\det (A)]^{-1}\left((-1)^2 \det \begin{bmatrix} 1 & -1 \\ 0 & 3 \end{bmatrix}, \ (-1)^3 \det \begin{bmatrix} 0 & -1 \\ 1 & 3 \end{bmatrix}, \right.$$

$$\left. (-1)^4 \det \begin{bmatrix} 0 & 1 \\ 1 & 0 \end{bmatrix}\right) = \tfrac{1}{2}(3, -1, -1).$$

The determinant can be used to evaluate the rank of a matrix. Once again the method is essentially useless as a practical means of computing the rank.

Theorem 6.6 *If* $A \in M_{m,n}(R)$ *and* $\rho(A) = k$, *then there exists a k-square submatrix of A whose determinant is nonzero. Moreover, if* $k < \min (m,n)$ *then every* $(k + t)$-*square submatrix of A is singular,* $k + t \leq \min (m,n)$.

Proof.   Let P be an m-square nonsingular matrix for which

$$PA = H$$

and H is the Hermite normal form for A. Since $\rho(A) = k$, we know that there exist integers $n_1 < n_2 < \cdots < n_k$ such that $H^{(n_i)}$ is the $m \times 1$ matrix with 1 as jth entry, all other entries 0. Then by the Cauchy-Binet Theorem (Theorem 6.1),

$$\begin{aligned} 1 &= \det (H[1, \ldots, k | n_1, \ldots, n_k]) \\ &= \det (PA[1, \ldots, k | n_1, \ldots, n_k]) \\ &= \sum_{\omega \in Q_{k,m}} \det (P[1, \ldots, k | \omega]) \det (A[\omega | n_1, \ldots, n_k]). \end{aligned}$$

Hence, not every one of the determinants $\det (A[\omega | n_1, \ldots, n_k])$ can be 0. Thus A possesses at least one nonzero k-square subdeterminant. On the other hand, let $L = P^{-1}$ so that

$$A = LH.$$

Let $\alpha \in Q_{k+1,m}, \beta \in Q_{k+1,n}$. Then once again from Theorem 6.1 we have

$$\det (A[\alpha | \beta]) = \sum_{\omega \in Q_{k+1,m}} \det (L[\alpha | \omega]) \det (H[\omega | \beta]). \tag{22}$$

Since H is the Hermite form of A and $\rho(A) = k$, we know that any submatrix with $k + 1$ rows must have a zero row (recall that H has only k nonzero rows). Thus every summand in (22) is 0 and det $(A[\alpha|\beta]) = 0$. From the Laplace Expansion Theorem (Theorem 6.2) we can conclude that once all $(k + 1)$-square subdeterminants are 0 then all $(k + t)$-square subdeterminants are also 0, $t \geq 1$.

As an example of an application of Theorem 6.6 we can prove

Theorem 6.7 *If U is an n-square unitary matrix and $1 \leq k \leq n$ and $\alpha \in Q_{k,n}$ then there exists $\beta \in Q_{k,n}$ such that*

$$\det (U[\alpha|\beta]) \neq 0. \tag{23}$$

Similarly, if $\alpha \in Q_{k,n}$ there exists $\gamma \in Q_{k,n}$ such that

$$\det (U[\gamma|\alpha]) \neq 0.$$

Proof.   Since U is unitary, rows numbered $\alpha_1, \ldots, \alpha_k$ are o.n. vectors and hence are linearly independent. This implies that the matrix $U[\alpha|1, \ldots, n]$ has rank k. We can then directly apply Theorem 6.6 to conclude (23). The second part of the theorem follows similarly.

The Cauchy-Binet theorem can also be used to prove the following interesting result.

Theorem 6.8 *If $A \in M_{m,n}(R)$, and R is the real-number field, then $\rho(A^T A) = \rho(A A^T) = \rho(A)$.*

Proof.   We verify that $\rho(A^T A) = \rho(A)$. Let $k = \rho(A)$. Then, if $\alpha, \beta \in Q_{k+1,m}$, we have by Theorem 6.1 and (15)

$$
\begin{aligned}
\det ((A^T A)[\alpha|\beta]) &= \sum_{\omega \in Q_{k+1,m}} \det (A^T[\alpha|\omega]) \det (A[\omega|\beta]) \\
&= \sum_{\omega \in Q_{k+1,m}} \det (A[\omega|\alpha]) \det (A[\omega|\beta]).
\end{aligned}
$$

Now Theorem 6.6 tells us that every summand in this last sum is zero because $\rho(A) = k$. Hence every $(k + 1)$-square subdeterminant of $A^T A$ is 0 and Theorem 6.6 implies that $\rho(A^T A) \leq k$. On the other hand, if

$\alpha \in Q_{k,m}$ then

$$\det ((A^T A)[\alpha|\alpha]) = \sum_{\omega \in Q_{k,m}} \det (A^T[\alpha|\omega]) \det (A[\omega|\alpha]) \qquad (24)$$

$$= \sum_{\omega \in Q_{k,m}} (\det (A[\omega|\alpha]))^2.$$

We know that some k-square subdeterminant of A is nonzero, say $A[\omega^0|\alpha^0]$. Take $\alpha = \alpha^0$ in (24), and then the sum on the right has the positive summand $(\det (A[\omega^0|\alpha^0]))^2$ and is therefore positive. It follows from Theorem 6.6 that $\rho(A^T A) \geq k$.

We can give a simpler proof that does not rely on the Cauchy-Binet theorem. But this shorter argument does not exhibit the very interesting identity (24). Let (x,y) denote the standard inner product for both $V_n(R)$ and $V_m(R)$. Now if $Ax = 0$, $x \in V_n(R)$, then clearly $A^T Ax = 0$. Conversely, if $A^T Ax = 0$, then $0 = (A^T Ax,x) = (Ax,Ax) = \|Ax\|^2$. Hence $Ax = 0$. It follows that A and $A^T A$ have the same null space (see Sect. 2.1) and hence have the same rank (Theorem 3.3).

Quiz

Answer true or false (in what follows R is a field):

1. If $A \in M_n(R)$ and R is the complex number field then $\det (AA^*) \geq 0$.
2. If A is an n-square matrix with positive entries then $\det (A) > 0$.
3. If R is the complex number field, $A \in M_n(R)$, and

$$B = \left[\begin{array}{c|c} 0 & A \\ \hline A^* & 0 \end{array}\right] \in M_{2n}(R)$$

then $\det (B) = (-1)^n |\det (A)|^2$.
4. If A is a real 3×4 matrix and

$$AA^T = \begin{bmatrix} 1 & 0 & 0 \\ 0 & 1 & 0 \\ 0 & 0 & 0 \end{bmatrix}$$

then $\rho(A) = 2$.
5. If

$$A = \begin{bmatrix} 0 & 0 & 1 & 0 \\ 0 & 0 & 0 & -1 \\ 2 & 3 & 4 & 5 \\ 6 & 7 & 8 & 9 \end{bmatrix}$$

then $\det (A) = 4$.

6. If $A \in M_n(R)$, $1 \leq k \leq n$, and for some $\alpha \in Q_{k,n}$ and every $\beta \in Q_{k,n}$ $A[\alpha|\beta]$ or $A(\alpha|\beta)$ has a zero row, then A is singular.

7. If $A \in M_n(R)$, $1 \leq k \leq n$, and for each $\alpha,\beta \in Q_{k,n}$ both $A[\alpha|\beta]$ and $A(\alpha|\beta)$ are nonsingular, then A is nonsingular.

8. If $A = A_1 \dotplus \cdots \dotplus A_k$ and each A_i is a square matrix then

$$\det (A) = \prod_{i=1}^{k} \det (A_i).$$

9. The third column of the adjugate of

$$A = \begin{bmatrix} 2 & 3 & 1 \\ 1 & 2 & 3 \\ 3 & 1 & 2 \end{bmatrix}$$

is $(\operatorname{adj} A)^{(3)} = (7,-5,1)$.

10. If U is an n-square unitary matrix, then $|\det (U)| = 1$.

Exercises

1. Using Theorem 6.1 and Theorem 6.2, give another proof of the fact that $\rho(AB) \leq \min \{\rho(A), \rho(B)\}$.

2. Show that if R is the field of complex numbers and $A \in M_{m,n}(R)$, then $\rho(A) = \rho(AA^*) = \rho(A^*A)$.

3. Show that if A is an n-square matrix with integer entries, then A is a unit (i.e., A^{-1} has integer entries), if and only if $\det (A) = \pm 1$.

4. Show that the product of unit matrices with integer entries is a unit matrix with integer entries.

5. Prove an analogue of the Laplace Expansion Theorem for permanents. That is, show that if $A \in M_n(R)$, $1 \leq k \leq n$, and $\alpha \in Q_{k,n}$ then

$$\operatorname{per} (A) = \sum_{\omega \in Q_{k,n}} \operatorname{per} (A[\alpha|\omega]) \operatorname{per} (A(\alpha|\omega)).$$

6. Let $A \in M_{m,n}(R)$, where R is the real number field. Prove that $\det (AA^T) \geq 0$ by using Theorem 6.1.

7. Let $A \in M_n(R)$, $1 \leq k \leq n$. Let $\alpha,\beta \in Q_{k,n}$. Prove that

$$\sum_{\omega \in Q_{k,n}} (-1)^{s(\alpha) + s(\omega)} \det (A[\alpha|\omega]) \det (A(\beta|\omega)) = \begin{cases} \det (A) & \text{if } \beta = \alpha, \\ 0 & \text{if } \beta \neq \alpha. \end{cases}$$

Characteristic Roots

3.1 Characteristic Roots and Characteristic Vectors

Much information about the effect of a linear transformation of a vector space into itself can be obtained by an examination of the vectors or the subspaces left invariant under the transformation. A vector v in a space V is said to be a *fixed* vector under a linear transformation $T \in L(V,V)$ if

$$T(v) = v.$$

Clearly, the zero vector is a fixed vector of any linear transformation. Also, if $v \neq 0$ is a fixed vector of T then any vector in the one-dimensional subspace $\langle v \rangle$ is a fixed vector. This is a special case of a subspace that is mapped into itself by a linear transformation. In general,

Definition 1.1 (*Invariant subspace*) *A subspace W of a vector space V is said to be **invariant** under a linear transformation $T \in L(V,V)$ if $T(u) \in W$ for every $u \in W$.*

In particular, a one-dimensional space $\langle u \rangle$, $u \neq 0$, is invariant under T if

$$T(u) = \lambda u$$

where λ is a scalar. For, if ξu is any vector of $\langle u \rangle$, then

$$T(\xi u) = \xi T(u) = \xi \lambda u = \lambda(\xi u).$$

Note that T maps every vector ξu of $\langle u \rangle$ into a fixed multiple of itself.

Definition 1.2 (*Characteristic roots and vectors*) *Let V be a finite-dimensional vector space with $T \in L(V,V)$. Suppose that u is a nonzero vector in V such that*

$$T(u) = \lambda u$$

*for some scalar λ. Then u is called a **characteristic vector** (or an **eigenvector** or a **proper vector** or a **latent vector**) of T and the number λ is the corresponding **characteristic root** (or **eigenvalue** or **proper value** or **latent root**) of T.*

In the present chapter we shall study characteristic roots and characteristic vectors of linear transformations and matrices.

Let $G = \{g_1, \ldots, g_n\}$ be a basis of a vector space V. Let $T \in L(V,V)$ and $A = [T]_G^G$. Suppose that u is a characteristic vector of T,

$$T(u) = \lambda u, \quad u \neq 0, \tag{1}$$

and that $x = (x_1, \ldots, x_n)$ is the coordinate vector for u relative to the basis G. Then, by Theorem 1.1 in Chap. 2,

$$Ax = \lambda x, \quad x \neq 0, \tag{2}$$

or

$$(\lambda I_n - A)x = 0, \quad x \neq 0. \tag{3}$$

Thus the problem of finding characteristic roots and vectors of a linear transformation T is reduced to that of finding the values λ for which the matrix $\lambda I_n - A$ has a nonzero vector in its null space, i.e., for which $\lambda I_n - A$ is singular and therefore

$$\det (\lambda I_n - A) = 0. \tag{4}$$

Using the language of linear equations, the problem is to find a number λ for which the homogeneous system

$$\sum_{j=1}^{n} (\lambda \delta_{ij} - a_{ij})x_j = 0, \quad i = 1, \ldots, n, \tag{5}$$

has nontrivial solutions, i.e., not all $x_j = 0, j = 1, \ldots, n$. Again, a necessary and sufficient condition for (5) to have such solutions is (4). When the expression on the left-hand side of (4) is expanded according to the definition of the determinant we can collect the various powers of λ together to obtain an expression

$$p(\lambda) = \lambda^n + c_{n-1}\lambda^{n-1} + \cdots + c_0.$$

The problem, which is a familiar one from high school algebra, is to find the value of the "unknown" λ which makes $p(\lambda)$ zero. The traditional way this is expressed is that we wish to solve the "polynomial equation"

$$p(\lambda) = 0.$$

To make sense of the above statement really requires a considerable effort which is beyond the purview of this book. However, we state the following definitions and results.

Definition 1.3 (Indeterminate) *Let R be a field and let R_1 be a commutative ring containing R such that the operations in R_1 are the same for elements of R as the operations in R. Suppose that $x_1, \ldots, x_n$ are elements of R_1. A **monomial in** $x_1, \ldots, x_n$ is an element of R_1 of the form $x_1^{m_1}x_2^{m_2} \ldots x_n^{m_n}$ in which $m_1, \ldots, m_n$ are non-negative integers. A **polynomial in** $x_1, \ldots, x_n$, denoted by $f(x_1, \ldots, x_n)$, is a sum of terms of the form $c_{m_1 \cdots m_n}x_1^{m_1} \cdots x_n^{m_n}, c_{m_1 \cdots m_n} \in R$, in which the summation is over distinct sequences $(m_1, \ldots, m_n)$:*

$$f(x_1, \ldots, x_n) = \Sigma c_{m_1 \cdots m_n}x_1^{m_1} \cdots x_n^{m_n}. \tag{6}$$

The elements $c_{m_1 \cdots m_n}$ are called **coefficients**. The elements $x_1, \ldots, x_n$ are said to be **independent indeterminates** (or in some contexts **unknowns**) over R if the only polynomial which is zero is the one for which all the coefficients are zero. If x_1 is a single indeterminate this definition means that whenever $c_n x_1{}^n + c_{n-1} x_1{}^{n-1} + \cdots + c_0 = 0$, $c_j \in R$, $j = 0, \ldots, n$, then $c_0 = c_1 = \cdots = c_n = 0$. The **degree** of a polynomial in the indeterminates $x_1, \ldots, x_n$, denoted by $\deg f(x_1, \ldots, x_n)$, is just the largest value of $m_1 + \cdots + m_n$ appearing in (6) for which the corresponding coefficient $c_{m_1 \cdots m_n}$ is not zero.

The most important results about this situation are described in the following two theorems. The first of these has been implicitly used since grade 9 whenever anyone says "let $x_1, \ldots, x_n$ be unknowns."

Theorem 1.1 *If R is a field and n a positive integer then there exists a ring R_1 containing R together with a set of n independent indeterminates over R.*

We do not prove Theorem 1.1 here but the result is familiar enough to be used by the student.

If $x_1, \ldots, x_n$ are the n independent indeterminates produced by Theorem 1.1 then the set of all polynomials in these indeterminates is denoted by $R[x_1, \ldots, x_n]$. With the familiar operations between polynomials, $R[x_1, \ldots, x_n]$ becomes a commutative ring. But more is true. We can define *rational forms* to be ordered pairs of polynomials $(f(x_1, \ldots, x_n), g(x_1, \ldots, x_n)) = (f,g)$, in which $g(x_1, \ldots, x_n) \neq 0$. Then if we define the product of two rational forms by $(f,g)(f_1,g_1) = (ff_1, gg_1)$ and the sum by $(f,g) + (f_1,g_1) = (g_1 f + f_1 g, gg_1)$, the set of all rational forms can easily be proved to be a field. Of course, it is customary from high school algebra to write (f,g) as $f(x_1, \ldots, x_n)/g(x_1, \ldots, x_n)$. The field of all rational forms is denoted by $R(x_1, \ldots, x_n)$ and it is clear that we can think of any polynomial $f(x_1, \ldots, x_n)$ as being the rational form $f(x_1, \ldots, x_n)/1$, i.e., $R[x_1, \ldots, x_n] \subset R(x_1, \ldots, x_n)$. The point of this last remark is that we can treat polynomials as elements of a field, e.g., we can apply the theorems on determinants to matrices with polynomial entries.

Definition 1.4 **(Root)** *If R is a field, x an indeterminate over R, and*

$$f(x) = \sum_{t=0}^{n} c_t x^t$$

*is a polynomial in x, then a number $r \in R$ is a **zero** of $f(x)$ if*

$$f(r) = \sum_{t=0}^{n} c_t r^t = 0.$$

*We also say that r is a **root** of the polynomial **equation***

$$\sum_{t=0}^{n} c_t x^t = 0.$$

In general, if $f(x_1, \ldots, x_n)$ is a polynomial in n independent indeterminates over R and $a_1, \ldots, a_n$ are elements of R then $f(a_1, \ldots, a_n)$ is the element of R obtained by replacing x_t by a_t, $t = 1, \ldots, n$, in $f(x_1, \ldots, x_n)$.

In a given field a polynomial equation may or may not have a root. On the other hand, the field of complex numbers has the following remarkable property known as the "fundamental theorem of algebra."

Theorem 1.2 *If R is the field of complex numbers, x is an indeterminate over R, and $f(x)$ is a polynomial in x of degree $n(n \geq 1)$, then there exist numbers $r_1, \ldots, r_n, c_n$ in R such that*

$$f(x) = c_n \prod_{i=1}^{n} (x - r_i).$$

Moreover, the numbers $r_1, \ldots, r_n$ (which need not be distinct) are exactly all the roots of the equation

$$f(x) = 0.$$

The proof of this theorem is beyond the scope of this book.

Henceforth the field R will be assumed to be the field of complex numbers unless otherwise stated.

Definition 1.5 *(Characteristic polynomial)* *Let $T \in L(V,V)$ and let $A = [T]_G^G$. Then the matrix $\lambda I_n - A$ is called the **characteristic matrix** of A, the polynomial*

$$\det (\lambda I_n - A)$$

is called the **characteristic polynomial** *of A* (*and of T*) *and its zeros*
$\lambda_1, \ldots, \lambda_n$ *are the* **characteristic roots** *of A* (*and of T*). *The equation*

$$\det (\lambda I_n - A) = 0$$

is called the **characteristic equation** *of A and any nonzero n-tuple x*
such that

$$(\lambda_t I_n - A)x = 0,$$

for some characteristic root λ_t, is called a **characteristic vector** *of A*
corresponding to λ_t.

In what follows we shall reserve the use of the letter λ to denote an
indeterminate. Characteristic roots will be denoted by adding a sub-
script: $\lambda_1, \ldots, \lambda_n, \lambda_i, \lambda_t$, etc.

It may appear from the above definition of the characteristic polynomial of a linear transformation that it depends on the matrix representation A. This, however, is not the case. For, by Theorem 5.4 in
Chap. I, if G and H are bases of $V, T \in L(V,V)$, $A = [T]_G{}^G$ and $B = [T]_H{}^H$, then A and B are similar. We now prove

Theorem 1.3 *Similar matrices have the same characteristic polynomial.*

Proof. Let $A = S^{-1}BS$. Then

$$\begin{aligned}
\det (\lambda I_n - A) &= \det (\lambda I_n - S^{-1}BS)\\
&= \det [S^{-1}(\lambda I_n - B)S]\\
&= \det (S^{-1}) \det (\lambda I_n - B) \det (S)\\
&= [\det (S)]^{-1} \det (\lambda I_n - B) \det (S)\\
&= \det (\lambda I_n - B).
\end{aligned}$$

The converse, however, is not true. Matrices with the same characteristic polynomial—i.e., with the same characteristic roots—need not
be similar. For example, both

$$\begin{bmatrix} 1 & 1 \\ 0 & 1 \end{bmatrix} \quad \text{and} \quad I_2 = \begin{bmatrix} 1 & 0 \\ 0 & 1 \end{bmatrix}$$

have the same characteristic polynomial $(\lambda - 1)^2$. However, for all nonsingular 2-square matrices S

$$S^{-1}I_2S = S^{-1}S = I_2$$

and thus $S^{-1}I_2S \neq \begin{bmatrix} 1 & 1 \\ 0 & 1 \end{bmatrix}$ for any S.

Also, in general, similar matrices can have different characteristic vectors corresponding to the same characteristic roots.

We proceed now to discuss the significance of the coefficients in the characteristic polynomial of a matrix.

Definition 1.6 (*Elementary symmetric functions*) *Let a_1, . . . , a_n be elements of the field R. Then the element of R defined by*

$$E_r(a_1, \ldots, a_n) = \sum_{(\omega_1, \ldots, \omega_r) \in Q_{r,n}} \prod_{i=1}^{r} a_{\omega_i}$$

is called the r-th elementary symmetric function of a_1, . . . , a_n.

For example,

$$\begin{aligned}
E_1(a_1,a_2,a_3,a_4) &= a_1 + a_2 + a_3 + a_4, \\
E_2(a_1,a_2,a_3,a_4) &= a_1a_2 + a_1a_3 + a_1a_4 + a_2a_3 + a_2a_4 + a_3a_4, \\
E_3(a_1,a_2,a_3,a_4) &= a_1a_2a_3 + a_1a_2a_4 + a_1a_3a_4 + a_2a_3a_4, \\
E_4(a_1,a_2,a_3,a_4) &= a_1a_2a_3a_4.
\end{aligned}$$

Here we are interested in the elementary symmetric functions of the characteristic roots of a complex matrix. For brevity, we shall denote the rth elementary symmetric function of the characteristic roots of A by $E_r(A)$ and call it the rth *elementary symmetric function* of A. The central result on the elementary symmetric functions of matrices is embodied in the following theorem.

Theorem 1.4   *If A is an n-square complex matrix and $1 \leq r \leq n$, then*

$$E_r(A) = \sum_{\omega \in Q_{r,n}} \det (A[\omega|\omega]).$$

In words: the r-th elementary symmetric function of the characteristic roots of A is equal to the sum of the determinants of all $\binom{n}{r}$ r-square principal submatrices of A.

Proof.   Consider the determinant of the matrix $B = X - A$, where $X = \mathrm{diag}\,(x_1, \ldots, x_n)$ and the x_j's are independent indeterminates over R. We shall first find the coefficient α_ω of $x_{\omega_1} \cdots x_{\omega_r}$, $\omega = (\omega_1, \ldots, \omega_r) \in Q_{r,n}$, in the expansion of det (B) and then equate the x_j's to an indeterminate λ. The terms involving $x_{\omega_1} \cdots x_{\omega_r}$ all occur in the $(n - r)!$ terms

$$\sum_{\sigma \in S_{n-r}} \epsilon(\sigma) b_{\omega_1,\omega_1} \cdots b_{\omega_r,\omega_r} b_{\tau_1,\tau_{\sigma(1)}} \cdots b_{\tau_{n-r},\tau_{\sigma(n-r)}}$$

in the expansion of det (B). Here $(\tau_1, \ldots, \tau_{n-r}) \in Q_{n-r,n}$ is complementary in $\{1, \ldots, n\}$ to $(\omega_1, \ldots, \omega_r)$ and σ is a permutation on $\{1, \ldots, n - r\}$. Now,

$$b_{\omega_k,\omega_k} = x_{\omega_k} - a_{\omega_k,\omega_k}, \quad k = 1, \ldots, r,$$

and

$$b_{\tau_s,\tau_{\sigma(s)}} = x_{\tau_s} \delta_{\tau_s,\tau_{\sigma(s)}} - a_{\tau_s,\tau_{\sigma(s)}}.$$

Thus every term involving precisely the product $x_{\omega_1} \cdots x_{\omega_r}$, i.e., no other x_j, is of the form

$$x_{\omega_1} \cdots x_{\omega_r}(-a_{\tau_1,\tau_{\sigma(1)}}) \cdots (-a_{\tau_{n-r},\tau_{\sigma(n-r)}}).$$

Moreover, all possible terms involving $x_{\omega_1} \cdots x_{\omega_r}$ are obtained by letting σ vary over all permutations on $\{1, \ldots, n - r\}$. Hence the coefficient α_ω of $x_{\omega_1} \cdots x_{\omega_r}$ is given by

$$\alpha_\omega = (-1)^{n-r} \sum_{\sigma \in S_{n-r}} \epsilon(\sigma) \prod_{i=1}^{n-r} a_{\tau_i,\tau_{\sigma(i)}} \tag{7}$$
$$= (-1)^{n-r} \det\,(A[\tau|\tau]).$$

Now if we set $x_1 = \cdots = x_n = \lambda$, then $x_{\omega_1} \cdots x_{\omega_r} = \lambda^r$ for any $\omega = (\omega_1, \ldots, \omega_r) \in Q_{r,n}$. Thus the coefficient of λ^r in det $(\lambda I_n - A) =$ det $(X - A)$ (when $x_1 = \cdots = x_n = \lambda$) is

$$(-1)^{n-r} \sum_{\tau \in Q_{n-r,n}} \det\,(A[\tau|\tau]). \tag{8}$$

For, each $\omega \in Q_{r,n}$ determines exactly one complementary sequence $\tau \in Q_{n-r,n}$. The coefficient of λ^r is obtained by summing all the α_ω. According to (7), this is the same thing as summing all the terms $(-1)^{n-r} \det (A[\tau|\tau])$, $\tau \in Q_{n-r,n}$. On the other hand, if $\lambda_1, \ldots, \lambda_n$ are the zeros of a polynomial

$$f(\lambda) = \sum_{r=0}^{n} c_r \lambda^r, \quad c_n = 1, \tag{9}$$

then, by Theorem 1.2,

$$f(\lambda) = \prod_{t=1}^{n} (\lambda - \lambda_t)$$

and we have

$$c_r = (-1)^{n-r} E_{n-r}(\lambda_1, \ldots, \lambda_n). \tag{10}$$

For, on multiplying out the product on the right-hand side of

$$f(\lambda) = \prod_{t=1}^{n} (\lambda - \lambda_t)$$

we can obtain all the terms in λ^r by taking the factor λ out of exactly r of the monomials $\lambda - \lambda_1, \ldots, \lambda - \lambda_n$ and the factors $-\lambda_t$ from the remaining $n - r$ monomials. Now set $f(\lambda) = \det (\lambda I_n - A)$ and compare (8) with (10). The result follows.

The elementary symmetric function $E_1(A)$ is given a special name.

Definition 1.7 (**Trace**)   *The elementary symmetric function* $E_1(A) = \lambda_1 + \cdots + \lambda_n$ *is called the **trace** of A and is denoted by* tr A.

Theorem 1.5   *If* $A = (a_{ij})$ *has characteristic roots* $\lambda_1, \ldots, \lambda_n$ *then*

(a) $\det (A) = \prod_{i=1}^{n} \lambda_i,$

(b)   tr $A = \sum_{i=1}^{n} a_{ii},$

(c) $E_r(S^{-1}AS) = E_r(A), \quad r = 1, \ldots, n.$

Proof. Parts (a) and (b) follow immediately from Theorem 1.4 and part (c) follows from (10) and Theorem 1.3.

In general, nothing can be said about matrices with equal traces beyond Theorem 1.5 (b). However, due to the linearity of the trace (see below), its invariance under similarity transformations, and the ease with which it can be evaluated, it plays an important role in matrix theory and applications. We give some properties of the trace.

Theorem 1.6   *If $A = (a_{ij})$ and $B = (b_{ij})$ are n-square complex matrices and α, β are any complex numbers, then*

(a) $\operatorname{tr}(\alpha A + \beta B) = \alpha \operatorname{tr} A + \beta \operatorname{tr} B,$

(b) $\operatorname{tr} A^* = \overline{\operatorname{tr} A},$

(c) $\operatorname{tr}(AB) = \operatorname{tr}(BA),$

(d) $\operatorname{tr}(AA^*) = \operatorname{tr}(A^*A) = \displaystyle\sum_{i,j=1}^{n} |a_{ij}|^2.$

Proof.

(a)
$$\operatorname{tr}(\alpha A + \beta B) = \sum_{i=1}^{n} (\alpha A + \beta B)_{ii}$$
$$= \alpha \sum_{i=1}^{n} a_{ii} + \beta \sum_{i=1}^{n} b_{ii}$$
$$= \alpha \operatorname{tr} A + \beta \operatorname{tr} B.$$

(b)
$$\operatorname{tr} A^* = \sum_{i=1}^{n} (\bar{A}^T)_{ii}$$
$$= \sum_{i=1}^{n} \overline{a_{ii}}$$
$$= \overline{\operatorname{tr} A}.$$

(c)
$$\operatorname{tr}(AB) = \sum_{i=1}^{n} (AB)_{ii}$$
$$= \sum_{i=1}^{n} \sum_{j=1}^{n} a_{ij} b_{ji}$$
$$= \sum_{j=1}^{n} \sum_{i=1}^{n} b_{ji} a_{ij}$$
$$= \sum_{j=1}^{n} (BA)_{jj}$$
$$= \operatorname{tr}(BA).$$

(d)
$$\text{tr } (AA^*) = \sum_{i=1}^{n} (AA^*)_{ii}$$

$$= \sum_{i=1}^{n} \sum_{j=1}^{n} (A)_{ij}(A^*)_{ji}$$

$$= \sum_{i=1}^{n} \sum_{j=1}^{n} a_{ij}\overline{a_{ij}}$$

$$= \sum_{i,j=1}^{n} |a_{ij}|^2.$$

Quiz

Answer true or false:

1. Two matrices are similar if and only if they have the same characteristic roots (including multiplicities).

2. The matrices $\begin{bmatrix} 2 & 5 & 0 \\ -2 & 1 & 4 \\ -1 & -3 & 3 \end{bmatrix}$ and $\begin{bmatrix} 0 & 5 & 2 \\ 4 & 1 & -2 \\ 3 & -3 & -1 \end{bmatrix}$ are similar.

3. If A is a 2-square matrix, then

$$\det (\lambda I_2 - A) = \lambda^2 - (\text{tr } A)\lambda + \det (A).$$

4. The number $\sqrt[3]{2}$ is an indeterminate over the field of rational numbers.

5. The matrices $\begin{bmatrix} 1 & 0 & 0 \\ 0 & 2 & 0 \\ 0 & 0 & 3 \end{bmatrix}$ and $\begin{bmatrix} 1 & 1 & 1 \\ 0 & 2 & 1 \\ 0 & 0 & 3 \end{bmatrix}$ have the same characteristic roots.

6. If x is a characteristic vector of A, then cx is also a characteristic vector of A for any scalar c.

7. Let A be an n-square matrix and let $\mu_1, \ldots, \mu_n$ be the characteristic roots of AA^*. Then $\mu_1 + \cdots + \mu_n = 0$ if and only if $A = 0$.·

8. If $D = \text{diag } (d_1, \ldots, d_n)$ then the numbers $d_1, \ldots, d_n$ are the characteristic roots of D and the vectors $e_1, \ldots, e_n$ are the corresponding characteristic vectors.

9. The determinant of a matrix A is zero if and only if A has a characteristic root equal to 0.

10. A linear transformation $T \in L(V,V)$ is singular if and only if a characteristic root of T is 0.

Exercises

1. Let S and T be in $L(V,V)$ and let S and T commute, i.e., $STv = TSv$ for all $v \in V$. Show that if u is a characteristic vector of T, not in the null space of S, then Su is a characteristic vector of T.

2. Suppose S and T are in $L(V,V)$ and have a characteristic vector in common:

$$Su = \lambda_i u, \quad Tu = \mu_j u, \quad u \in V, \quad u \neq 0.$$

Show that $\lambda_i \mu_j$ is a characteristic root of ST and of TS and that u is a corresponding characteristic vector. Deduce a theorem about characteristic roots and vectors of S^k.

3. Show that if λ_j is a characteristic root of $T \in L(V,V)$ and c is a nonzero scalar then $c\lambda_j$ is a characteristic root of $cT \in L(V,V)$.

4. Show that if $\lambda_1, \ldots, \lambda_n$ are the characteristic roots of a matrix A and c is any scalar then the characteristic roots of $cI_n + A$ are $c + \lambda_1, \ldots, c + \lambda_n$.

5. Let $T \in L(V_2(\text{Re}), V_2(\text{Re}))$ be defined by

$$T(v) = (v_1 - v_2, v_1 + v_2)$$

for all $v = (v_1, v_2) \in V_2(\text{Re})$. Show that T has no (real) characteristic roots or vectors.

6. Find the characteristic roots and vectors of the matrix

$$A = \begin{bmatrix} 4 & -3 & -2 \\ 2 & -1 & -2 \\ 3 & -3 & -1 \end{bmatrix}$$

Compute the determinants of all principal submatrices of A and verify Theorem 1.4.

7. Show that the matrix

$$B = \begin{bmatrix} 1 & 1 & -1 \\ -1 & 3 & -1 \\ -1 & 1 & 1 \end{bmatrix}$$

has a double characteristic root and three linearly independent characteristic vectors.

8. Show that the matrix

$$C = \begin{bmatrix} 3 & 2 & -6 \\ 0 & 2 & -1 \\ 1 & 1 & -2 \end{bmatrix}$$

has a triple characteristic root and that there is only one one-dimensional invariant subspace $\langle u \rangle$ of $V_3(R)$ such that $Cu \in \langle u \rangle$.

9. Show that if $A \in M_n(R)$ then A and A^T have the same characteristic roots, $\lambda_1, \ldots, \lambda_n$. Also show that the characteristic roots of cA are $c\lambda_i$, $i = 1, \ldots, n$.

3.2 Schur's Theorem

The evaluation of the characteristic roots of a given n-square matrix requires computation of an nth order determinant and the roots of an equation of nth degree—formidable, often impossible, tasks. There is, however, an important class of matrices whose characteristic roots are actually exhibited as certain entries of the matrix. In what follows we shall assume R is an arbitrary field.

Definition 2.1 (Triangular matrix) *A matrix $A \in M_n(R)$ is called **upper triangular** if $a_{ij} = 0$ whenever $i > j$. It is called **lower triangular** if $a_{ij} = 0$ whenever $i < j$.*

Clearly if A is upper or lower triangular then all its diagonals, except possibly the main diagonal, contain zero elements and

$$\det (A) = \prod_{i=1}^{n} a_{ii}. \tag{1}$$

If A is upper (lower) triangular then its characteristic matrix $\lambda I_n - A$ is upper (lower) triangular and therefore

$$\det (\lambda I_n - A) = \prod_{i=1}^{n} (\lambda - a_{ii}).$$

It follows that the characteristic roots of a triangular matrix are precisely the main diagonal entries of the matrix.

We have already encountered very special types of triangular matrices: the diagonal matrices. We have also seen in the preceding section an example of a matrix, i.e., $\begin{bmatrix} 1 & 1 \\ 0 & 1 \end{bmatrix}$, that is not similar to a diagonal matrix. Matrices that are similar to diagonal matrices have a simple structure and in general are easy to deal with.

Theorem 2.1 *A matrix $A \in M_n(R)$ is similar over R to a diagonal matrix in $M_n(R)$ if and only if A has n linearly independent characteristic vectors in $V_n(R)$. In fact, $S^{-1}AS$ is diagonal if and only if the columns of S are characteristic vectors of A.*

Proof. Recall that $(BC)^{(j)} = BC^{(j)}$ for any matrices B and C, that $B^{(j)} = Be_j$, and that, in particular, $[\text{diag }(\lambda_1, \ldots, \lambda_n)]^{(j)} = \lambda_j e_j$. Now suppose that A is similar to a diagonal matrix, i.e., that

$$S^{-1}AS = D = \text{diag }(\lambda_1, \ldots, \lambda_n)$$

for some $S \in M_n(R)$. Then

$$AS = SD$$

and

$$(AS)^{(j)} = (SD)^{(j)}, \quad j = 1, \ldots, n,$$

i.e.,

$$AS^{(j)} = S\lambda_j e_j = \lambda_j S^{(j)} \qquad (2)$$

Since S is nonsingular, its columns are linearly independent and, by (2), they are characteristic vectors of A.

To prove the converse, suppose that $v_1, \ldots, v_n$ in $V_n(R)$ are linearly independent characteristic vectors of A,

$$Av_j = \lambda_j v_j, \quad j = 1, \ldots, n. \qquad (3)$$

Let S be the n-square matrix whose jth column vector is v_j. Then

$$AS^{(j)} = \lambda_j S^{(j)}, \quad j = 1, \ldots, n.$$

The columns of S are linearly independent and hence S is nonsingular. [Chap. 2, Theorem 3.2(b).] We can reverse the steps in the first part of the proof to obtain

$$S^{-1}AS = \text{diag }(\lambda_1, \ldots, \lambda_n).$$

We now prove a theorem which will enable us to deduce an important sufficient condition for a matrix to be similar to a diagonal matrix.

Theorem 2.2 *Characteristic vectors corresponding to distinct characteristic roots of a matrix are linearly independent.*

Proof. Let $\lambda_1, \ldots, \lambda_k$ be distinct characteristic roots of an n-square matrix A and let $v_1, \ldots, v_k$ be the corresponding characteristic vectors. We shall prove by induction on k that $v_1, \ldots, v_k$ are linearly independent. The theorem holds trivially for $k = 1$. Assume that it is true for $k - 1$ distinct characteristic roots $(k > 1)$, and suppose that for scalars $c_i \in R$, $i = 1, \ldots, k$,

$$c_1 v_1 + \cdots + c_{k-1} v_{k-1} + c_k v_k = 0. \tag{4}$$

Then

$$c_1 A v_1 + \cdots + c_{k-1} A v_{k-1} + c_k A v_k = 0$$

or, since $v_1, \ldots, v_k$ are characteristic vectors of A corresponding to $\lambda_1, \ldots, \lambda_k$,

$$c_1 \lambda_1 v_1 + \cdots + c_{k-1} \lambda_{k-1} v_{k-1} + c_k \lambda_k v_k = 0. \tag{5}$$

Eliminate v_k between (4) and (5) to obtain

$$c_1 (\lambda_1 - \lambda_k) v_1 + \cdots + c_{k-1} (\lambda_{k-1} - \lambda_k) v_{k-1} = 0.$$

Now $v_1, \ldots, v_{k-1}$ are linearly independent and $\lambda_j - \lambda_k \neq 0$ for $j = 1, \ldots, k - 1$. Thus

$$c_1 = \cdots = c_{k-1} = 0.$$

Hence $c_k v_k = 0$ and, since v_k is a characteristic vector, we must have $c_k = 0$. The characteristic vectors $v_1, \ldots, v_k$ are therefore linearly independent.

Theorem 2.3 *If a matrix $A \in M_n(R)$ has n distinct characteristic roots then A is similar to a diagonal matrix.*

This result follows immediately from Theorems 2.1 and 2.2.

We saw that not every square matrix is similar to a diagonal matrix. We shall prove, however, a beautiful result due to I. Schur (Theorem 2.6), that every square complex matrix is similar, in a rather special way, to a triangular matrix. Although general triangular matrices do not possess the simple properties of diagonal matrices they have some remarkable algebraic features which make Schur's theorem of fundamental importance.

Theorem 2.4 *If $A = (a_{ij})$ and $B = (b_{ij})$ are upper triangular matrices in $M_n(R)$ then $S = (s_{ij}) = A + B$ and $P = (p_{ij}) = AB$ are also upper* *triangular. Moreover,*

$$s_{ii} = a_{ii} + b_{ii} \quad and \quad p_{ii} = a_{ii}b_{ii}, \quad i = 1, \ldots, n.$$

Proof. Since A and B are upper triangular, $a_{ij} = 0$ if $i > j$ and $b_{ij} = 0$ if $i > j$. Therefore $s_{ij} = a_{ij} + b_{ij} = 0$ if $i > j$ and thus S is upper triangular. Also, if $i > j$ then for any integer k either $i > k$ or $k > j$ and thus either $a_{ik} = 0$ or $b_{kj} = 0$. Hence, for $i > j$,

$$p_{ij} = \sum_{k=1}^{n} a_{ik}b_{kj} = 0$$

and P is upper triangular. Now, $s_{ii} = a_{ii} + b_{ii}$, by the definition of matrix addition. Also,

$$p_{ii} = \sum_{k=1}^{n} a_{ik}b_{ki} = a_{ii}b_{ii}$$

since $a_{ik} = 0$ for $i > k$ and $b_{ki} = 0$ for $i < k$.

Theorem 2.5 *If A, B, S and P are the matrices defined in Theorem 2.4 then their characteristic roots are $a_{ii}, b_{ii}, a_{ii} + b_{ii}, a_{ii}b_{ii}, i = 1, \ldots, n,$ respectively.*

Clearly analogous results hold for lower triangular matrices. To avoid repeating this kind of statement and to simplify our terminology we shall use henceforth the term "triangular" to mean "upper triangular."

Definition 2.2 (*Unitary similarity*) *Two complex matrices A and B are said to be **unitarily similar** (or **unitarily equivalent**) if there exists a unitary matrix U such that $U^*AU = B$. Two real matrices A and B are called **orthogonally similar** if there exists a (real) orthogonal matrix P such that $P^TAP = B$.*

Recall that, for unitary matrices, $U^* = U^{-1}$ and hence the term unitary similarity is an appropriate one.

Theorem 2.6 *Every n-square complex matrix is unitarily similar to a triangular matrix. Every real matrix with real characteristic roots is orthogonally similar to a real triangular matrix.*

Proof.   Let $A \in M_n(R)$ where R is the complex field. We prove by induction on n that there exists a unitary matrix U such that $U^*AU = B$ is triangular. Let λ_1 be a characteristic root of A and x_1 a corresponding characteristic vector of unit length. Construct a unitary matrix S such that $S^{(1)} = x_1$. This can be accomplished in the following way. Complete x_1 to a basis of $V_n(R)$,

$$\langle x_1, y_2, \ldots, y_n \rangle = V_n(R).$$

(If the kth entry in x_1 is nonzero, then the $y_2, \ldots, y_n$ can be chosen, e.g., to be $e_1, \ldots, e_{k-1}, e_{k+1}, \ldots, e_n$.) Then use the Gram-Schmidt process in $V_n(R)$ with the standard inner product to construct an o.n. set $x_1, x_2, \ldots, x_n$. Let S be the matrix in $M_n(R)$ whose jth column is x_j. We assert that S is unitary. For

$$\begin{aligned} (S^*S)_{ij} &= \sum_{k=1}^{n} (S^*)_{ik} S_{kj} \\ &= \sum_{k=1}^{n} S_{kj} \bar{S}_{ki} \\ &= (x_j, x_i) \\ &= \delta_{ij}. \end{aligned}$$

In other words, $S^*S = I_n$. We compute

$$\begin{aligned} (S^{-1}AS)^{(1)} &= S^{-1}AS^{(1)} \\ &= S^{-1}Ax_1 \\ &= S^{-1}\lambda_1 x_1 \\ &= \lambda_1 S^{-1} S^{(1)} \\ &= \lambda_1 (S^{-1}S)^{(1)} \\ &= \lambda_1 (I_n)^{(1)} \\ &= \lambda_1 e_1 \end{aligned}$$

and thus

$$S^*AS = \begin{bmatrix} \lambda_1 & z \\ 0 & \\ \cdot & \\ \cdot & A_1 \\ \cdot & \\ 0 & \end{bmatrix}$$

where z represents a $1 \times (n-1)$ submatrix and A_1 is an $(n-1)$-square submatrix of S^*AS. Therefore, by the induction hypothesis, there exists a unitary $(n-1)$-square matrix L_1 such that $L_1^*A_1L_1$ is triangular. Let $L = [1] \dotplus L_1$ and observe that L is unitary and $L^* = [1] \dotplus L_1^*$. Thus

$$L^*(S^*AS)L = \begin{bmatrix} \lambda_1 & \vdots & y \\ 0 & & \\ \cdot & \vdots & L_1^*A_1L_1 \\ \cdot & & \\ \cdot & \vdots & \\ 0 & \vdots & \end{bmatrix}$$

is triangular. Now set $U = SL$. Then U is unitary (see Exercise 5, Sec. 1.4) and U^*AU is triangular.

If A and λ_1 are real, then x_1 can be chosen to be real. For if $x_1 = (x_{1,1}, \ldots, x_{1,n})$ then the real homogeneous system

$$(\lambda_1 I_n - A)x_1 = 0$$

always has a nontrivial real solution (see Chap. 2, Theorems 3.3 and 5.3). Now, the Gram-Schmidt process involves only rational operations and extractions of square roots of positive real numbers. Thus S can be chosen to be real orthogonal. If the other characteristic roots of A are also real then, by the induction hypothesis, L_1 can be chosen to be real orthogonal and therefore $P = SL$ is real orthogonal and P^TAP is real triangular.

The method of proof in the above theorem is essentially constructive. We shall illustrate it in the following example. Let

$$A = \begin{bmatrix} 2 & 1 & 0 & -1 \\ -1 & 1 & 1 & 1 \\ 0 & 1 & 1 & 0 \\ 1 & 1 & 0 & 0 \end{bmatrix}.$$

The problem is to find a unitary matrix U such that U^*AU is triangular. Since all the row sums of A are equal to 2,

$$Au = 2u$$

where $u = (1,1,1,1)$. Therefore u is a characteristic vector of A. Let

$x_1 = u/\|u\| = \frac{1}{2}(1,1,1,1)$ and let S be the orthogonal matrix

$$\frac{1}{2}\begin{bmatrix} 1 & 1 & 1 & 1 \\ 1 & 1 & -1 & -1 \\ 1 & -1 & 1 & -1 \\ 1 & -1 & -1 & 1 \end{bmatrix}.$$

Then

$$S^*AS = \begin{bmatrix} 2 & 1 & 0 & -1 \\ 0 & 0 & 0 & 0 \\ 0 & 1 & 1 & 0 \\ 0 & 2 & 1 & 1 \end{bmatrix}.$$

Let

$$A_1 = \begin{bmatrix} 0 & 0 & 0 \\ 1 & 1 & 0 \\ 2 & 1 & 1 \end{bmatrix}.$$

Clearly 1 is a characteristic root of A_1. A corresponding unit characteristic vector is $(0,0,1)$. Let L_1 be the orthogonal matrix

$$\begin{bmatrix} 0 & -\frac{4}{5} & \frac{3}{5} \\ 0 & \frac{3}{5} & \frac{4}{5} \\ 1 & 0 & 0 \end{bmatrix}.$$

(It would be easy to choose a matrix that would accelerate the triangularization. However, the purpose of this example is to illustrate the method of proof in Theorem 2.6.) Then

$$L_1^T A_1 L_1 = \begin{bmatrix} 1 & -1 & 2 \\ 0 & -\frac{3}{25} & \frac{21}{25} \\ 0 & -\frac{4}{25} & \frac{28}{25} \end{bmatrix}.$$

The matrix

$$A_2 = \frac{1}{25}\begin{bmatrix} -3 & 21 \\ -4 & 28 \end{bmatrix}$$

is clearly singular, and thus 0 is a characteristic root and the corresponding unit characteristic vector is $\dfrac{1}{\sqrt{50}}$ (7   1). Let L_2 be the orthog-

onal matrix

$$\frac{1}{\sqrt{50}}\begin{bmatrix} 7 & -1 \\ 1 & 7 \end{bmatrix}.$$

Now let

$$U = S(I_1 \dotplus L_1)(I_2 \dotplus L_2) = \begin{bmatrix} \frac{1}{2} & \frac{1}{2} & 0 & \frac{1}{\sqrt{2}} \\[2mm] \frac{1}{2} & -\frac{1}{2} & -\frac{1}{\sqrt{2}} & 0 \\[2mm] \frac{1}{2} & -\frac{1}{2} & \frac{1}{\sqrt{2}} & 0 \\[2mm] \frac{1}{2} & \frac{1}{2} & 0 & -\frac{1}{\sqrt{2}} \end{bmatrix}$$

Then U is orthogonal and

$$U^T A U = \begin{bmatrix} 2 & -1 & -\dfrac{1}{\sqrt{2}} & \dfrac{1}{\sqrt{2}} \\[2mm] 0 & 1 & -\dfrac{1}{\sqrt{2}} & \dfrac{3}{\sqrt{2}} \\[2mm] 0 & 0 & 0 & 1 \\[2mm] 0 & 0 & 0 & 1 \end{bmatrix}$$

is triangular.

Definition 2.3   (*Matrix polynomial*)   *If $p(x)$ is a polynomial in x,*

$$p(x) = \sum_{i=0}^{m} c_i x^i, \quad c_i \in R,$$

and $A \in M_n(R)$, then $p(A)$ is defined by

$$p(A) = \sum_{i=0}^{m} c_i A^i,$$

*where A^0 is interpreted as I_n. If $p(A) = 0$, then A is said to **satisfy** the equation $p(x) = 0$.*

The following result is an important consequence of Theorem 2.6.

Theorem 2.7 *If $p(x)$ is a polynomial in an indeterminate x and A an n-square matrix with characteristic roots λ_i, $i = 1, \ldots, n$, then the characteristic roots of $p(A)$ are $p(\lambda_i)$, $i = 1, \ldots, n$.*

Proof. Let $p(x) = \sum_{i=0}^{m} c_i x^i$ and let U be a unitary matrix such that $U^*AU = B$ is triangular. Note that $(U^*AU)^i = U^*A^iU$ for any i (i.e., $U^*AUU^*AU = U^*A^2U$, etc.). Therefore

$$U^*p(A)U = \sum_{i=0}^{m} c_i B^i$$

is triangular and, by Theorem 2.5, its characteristic roots are

$$\sum_{i=0}^{n} c_i \mu_j{}^i, \quad j = 1, \ldots, n,$$

where the μ_j's are the characteristic roots of B. But B and A are similar and thus, by Theorem 1.3, $\{\mu_1, \ldots, \mu_n\} = \{\lambda_1, \ldots, \lambda_n\}$.

The following result is known as the *Cayley-Hamilton Theorem.*

Theorem 2.8 *Every matrix $A \in M_n(R)$ satisfies its characteristic equation.*

Proof. Let $f(\lambda) = \lambda^n + c_{n-1}\lambda^{n-1} + \cdots + c_1\lambda + c_0$ be the characteristic polynomial of A. Then, by Theorem 6.4, Chap. 2,

$$(\lambda I_n - A) \text{ adj } (\lambda I_n - A) = \det (\lambda I_n - A)I_n = f(\lambda)I_n. \qquad (6)$$

Now, adj $(\lambda I_n - A)$ is a matrix whose entries are determinants (multiplied by ± 1) of $(n-1)$-square submatrices of $\lambda I_n - A$. Hence adj $(\lambda I_n - A)$ is a matrix whose entries are polynomials of degree not exceeding $n-1$. Thus adj $(\lambda I_n - A) = B_{n-1}\lambda^{n-1} + \cdots + B_1\lambda + B_0$ where B_j are matrices with scalar entries. Hence (6) can be written

$$(\lambda I_n - A)(B_{n-1}\lambda^{n-1} + \cdots + B_1\lambda + B_0)$$
$$= (\lambda^n + c_{n-1}\lambda^{n-1} + \cdots + c_1\lambda + c_0)I_n. \qquad (7)$$

Comparing the matrix coefficients of λ^n, . . . , λ, λ^0 on both sides of (7), we obtain

$$
\begin{aligned}
B_{n-1} &= I_n \\
B_{n-2} - AB_{n-1} &= c_{n-1}I_n \\
B_{n-3} - AB_{n-2} &= c_{n-2}I_n \\
\cdots \cdots \cdots \cdots \cdots \cdots & \\
B_0 - AB_1 &= c_1 I_n \\
- AB_0 &= c_0 I_n
\end{aligned}
$$

Multiplying these equalities by A^n, A^{n-1}, . . . , A, I_n respectively and adding them up we obtain

$$
0 = A^n + c_{n-1}A^{n-1} + \cdots + c_1 A + c_0 I_n = f(A).
$$

Quiz

Answer true or false:

1. If λ_1, . . . , λ_n and μ_1, . . . , μ_n are the characteristic roots of A and B respectively then $\lambda_1 + \mu_1$, . . . , $\lambda_n + \mu_n$ are the characteristic roots of $A + B$.

2. The matrices $\begin{bmatrix} 1 & 0 & 0 \\ 1 & 2 & 0 \\ 1 & 1 & 3 \end{bmatrix}$ and $\begin{bmatrix} 1 & 1 & 0 \\ 0 & 2 & 1 \\ 0 & 0 & 3 \end{bmatrix}$ are similar.

3. The matrix $\begin{bmatrix} 2 & 0 & 0 \\ 0 & 2 & 1 \\ 0 & 0 & 1 \end{bmatrix}$ is similar to a diagonal matrix.

4. If all the characteristic roots of A are 0, then $A = 0$.

5. If $A = \text{diag } (d_1, \ldots, d_n)$ and $B = \text{diag } (d_{\sigma(1)}, \ldots, d_{\sigma(n)})$, where σ is a permutation in S_n, then A and B are similar.

    (In Questions 6, 7, and 8 an n-square *permutation matrix* P is one which has precisely one 1 and $n - 1$ zeros in each row and in each column. In other words, $p_{ij} = \delta_{i\sigma(j)}, i, j = 1, \ldots, n$, for some $\sigma \in S_n$.)

6. Two diagonal n-square matrices D_1 and D_2 are similar if and only if there exists a permutation matrix P such that $P^T D_1 P = D_2$.

7. Permutation matrices are unitary.

8. If two permutation matrices are similar they are equal.

9. A triangular matrix is similar to a diagonal matrix if and only if it has distinct characteristic roots.

10. If $A \in M_n(R)$ then there exists a polynomial $p(x)$ of degree not exceeding $n - 1$ such that $A^n = p(A)$.

Exercises

1. Show that the characteristic roots of the matrix

$$A = \begin{bmatrix} 1 & 0 & -2 \\ -1 & 1 & 0 \\ 0 & -2 & 4 \end{bmatrix}$$

are all real. Find an orthogonal matrix P such that $P^T A P$ is triangular.

2. Show that the matrix A in Exercise 1 is not similar to a diagonal matrix.

3. Prove that (upper) triangular n-square matrices form a ring.

4. Prove that if all the characteristic roots of a matrix A are 0 then A is *nilpotent*, i.e., $A^k = 0$ for some integer k.

5. Show that if a matrix A is *idempotent*, i.e., $A^2 = A$, then all its characteristic roots are either equal to 0 or 1.

6. Let $T \in L(V,V)$. Show that there exists an orthonormal basis of V such that $[T]_G{}^G$ is triangular.

3.3   Normal Matrices

In Sec. 1.4 we defined a normal matrix as one that commutes with its conjugate transpose. We there obtained some elementary results about normal matrices and normal transformations. We are now in a position to prove the main theorem on the structure of normal matrices and to deduce some remarkable properties of hermitian, skew-hermitian, and unitary matrices.

As we pointed out in the preceding section, n-square matrices that are similar to diagonal matrices possess characteristic vectors which form a basis of $V_n(R)$. Normal matrices are not only endowed with this important property but, as we shall see, they possess characteristic vectors that form an orthonormal basis for $V_n(R)$. This feature and its consequences make normal matrices particularly "well-behaved." Numerous results on normal matrices have been obtained. Some of these are either untrue or still undecided for general matrices. The importance of normal matrices is further enhanced by the fact that many applications in mathematics and physics involve only normal matrices.

In what follows R will denote the complex number field.

Theorem 3.1   *A matrix is normal if and only if it is unitarily similar to a diagonal matrix.*

Proof.   If $A = UDU^*$, where D is diagonal and U is unitary then $A^* = UD^*U^* = U\bar{D}U^*$ and

$$
\begin{aligned}
AA^* &= (UDU^*)(U\bar{D}U^*) \\
&= UD\bar{D}U^* \\
&= U\bar{D}DU^* \\
&= (U\bar{D}U^*)(UDU^*) \\
&= A^*A.
\end{aligned}
$$

To prove the necessity, we first show that every normal triangular matrix is diagonal, and then apply Theorem 2.6. Suppose then that a matrix $B = (b_{ij})$ is both normal and upper triangular, i.e., that

$$
b_{ij} = 0 \quad \text{for } i > j
$$

and

$$
BB^* = B^*B.
$$

We use induction on n to show that B is diagonal.

Now

$$
(BB^*)_{1,1} = \sum_{j=1}^{n} |b_{1j}|^2
$$

while

$$
(B^*B)_{1,1} = \sum_{i=1}^{n} |b_{i1}|^2 \\
= |b_{11}|^2,
$$

since $b_{i1} = 0$ for $i > 1$.

Hence

$$
b_{1j} = 0, \quad j > 1,
$$

and

$$
B = [b_{11}] \dotplus B_1
$$

where $B_1 \in M_{n-1}(R)$ is upper triangular.

But

$$BB^* = [|b_{11}|^2] + B_1 B_1^*$$

and

$$B^*B = [|b_{11}|^2] + B_1^* B_1.$$

Thus B_1 is normal and therefore, by the induction hypothesis, it is diagonal. Hence B is diagonal.

Now, suppose that A is a normal matrix. By Theorem 2.6, there exists a unitary matrix U such that

$$U^*AU = B$$

where B is triangular. But

$$\begin{aligned}
BB^* &= (U^*AU)(U^*A^*U) \\
&= U^*AA^*U \\
&= U^*A^*AU \\
&= (U^*A^*U)(U^*AU) \\
&= B^*B,
\end{aligned}$$

and thus $B = U^*AU$ is diagonal.

The following important results follow directly from Theorem 3.1.

Theorem 3.2 *Let V be a unitary space and $T \in L(V,V)$. Then there exists an orthonormal basis G such that $[T]_G^G$ is diagonal if and only if T is normal.*

Theorem 3.3 *An n-square matrix is normal if and only if it possesses n orthonormal characteristic vectors.*

The first result follows from Ex. 6, Sec. 3.2, and Theorem 3.1, and the other from Theorems 2.1 and 3.1.

Theorem 3.4 *The rank of a normal n-square matrix with exactly p zero characteristic roots is $n - p$.*

This property does not hold in general for nonnormal matrices. For example, the 2-square matrix $\begin{bmatrix} 0 & 1 \\ 0 & 0 \end{bmatrix}$ has 2 zero characteristic roots but its rank is 1, not 0.

Other important consequences of Theorem 3.1 are the following alternative characterizations of hermitian, skew-hermitian and unitary matrices.

Theorem 3.5

(a) *A normal matrix is hermitian if and only if its characteristic roots are real.*

(b) *A normal matrix is skew-hermitian if and only if its characteristic roots are pure imaginary.*

(c) *A normal matrix is unitary if and only if its characteristic roots are of modulus 1.*

Proof. If A is normal then, by Theorem 3.1, there exists a unitary matrix U such that

$$U^*AU = \text{diag}(\lambda_1, \ldots, \lambda_n)$$

where $\lambda_1, \ldots, \lambda_n$ are the characteristic roots of U^*AU and therefore of A. All three parts of the theorem follow immediately from the following remarks.

If A is normal (hermitian, skew-hermitian, unitary) and U is unitary then U^*AU has the same property. For example,

$$\begin{aligned}
(U^*AU)(U^*AU)^* &= U^*AUU^*A^*U \\
&= U^*AA^*U \\
&= U^*A^*AU \\
&= (U^*A^*U)(U^*AU) \\
&= (U^*AU)^*(U^*AU).
\end{aligned}$$

Now,

(a) $\text{diag}(\lambda_1, \ldots, \lambda_n)$ is hermitian if and only if $\lambda_1, \ldots, \lambda_n$ are real;

(b) $\text{diag}(\lambda_1, \ldots, \lambda_n)$ is skew-hermitian if and only if $\lambda_1, \ldots, \lambda_n$ are pure imaginary;

(c) $\text{diag}(\lambda_1, \ldots, \lambda_n)$ is unitary if and only if $|\lambda_1| = \cdots = |\lambda_n| = 1$.

The following celebrated inequality is due to I. Schur and gives us further information about the relationship between the characteristic roots and the entries of a matrix.

Theorem 3.6 *Let $A = (a_{ij})$ be an n-square complex matrix with characteristic roots $\lambda_1, \ldots, \lambda_n$. Then*

$$\sum_{t=1}^{n} |\lambda_t|^2 \leq \sum_{i,j=1}^{n} |a_{ij}|^2. \qquad (1)$$

Equality holds in (1) if and only if A is normal.

Proof. By Theorem 2.6, there exists a unitary matrix U such that $U^*AU = B$ is upper triangular. Then

$$BB^* = (U^*AU)(U^*AU)^* = U^*AA^*U.$$

Thus BB^* and AA^* are similar and, by Theorem 1.5,

$$\text{tr } (BB^*) = \text{tr } (AA^*). \qquad (2)$$

Now, if $X = (x_{ij})$ is any n-square complex matrix

$$(XX^*)_{ii} = \sum_{j=1}^{n} x_{ij}\bar{x}_{ij} = \sum_{j=1}^{n} |x_{ij}|^2$$

and therefore, tr $(XX^*) = \sum_{i=1}^{n}\sum_{j=1}^{n} |x_{ij}|^2$. (See Theorem 1.6 (d).) Thus (2) becomes

$$\sum_{i,j=1}^{n} |b_{ij}|^2 = \sum_{i,j=1}^{n} |a_{ij}|^2.$$

But $b_{ij} = 0$ for $i > j$ and $b_{ii} = \lambda_i$, $i = 1, \ldots, n$, because A and B have the same characteristic roots. Therefore

$$\sum_{i,j=1}^{n} |b_{ij}|^2 = \sum_{i=1}^{n} |\lambda_i|^2 + \sum_{i<j} |b_{ij}|^2$$

and

$$\sum_{i=1}^{n} |\lambda_i|^2 + \sum_{i<j} |b_{ij}|^2 = \sum_{i,j=1}^{n} |a_{ij}|^2.$$

It follows that

$$\sum_{i=1}^{n} |\lambda_i|^2 \leq \sum_{i,j=1}^{n} |a_{ij}|^2$$

with equality if and only if $\sum_{i<j} |b_{ij}|^2 = 0$, i.e., if and only if B is diagonal or, by Theorem 3.1, if and only if A is normal.

Theorem 3.7 *If $\lambda_1, \ldots, \lambda_n$ are the characteristic roots of $A = (a_{ij})$ and $\mu = \max_{i,j} |a_{ij}|$, then*

(a) $|\det A| \leq n^{n/2}\mu^n,$
(b) $|\lambda_t| \leq n\mu, \quad t = 1, \ldots, n.$

Proof.   By Theorem 3.6,

$$\sum_{t=1}^{n} |\lambda_t|^2 \leq \sum_{i,j=1}^{n} |a_{ij}|^2.$$

Now, $|a_{ij}| \leq \mu$, $i,j = 1, \ldots, n$, and therefore

$$\sum_{t=1}^{n} |\lambda_t|^2 \leq n^2\mu^2. \tag{3}$$

Apply the arithmetic-geometric mean inequality and Theorem 1.5 to the numbers $|\lambda_1|^2, \ldots, |\lambda_n|^2$ to obtain

$$\frac{1}{n}\sum_{t=1}^{n} |\lambda_t|^2 \geq \left(\prod_{t=1}^{n} |\lambda_t|^2\right)^{1/n} = |\det A|^{2/n} \tag{4}$$

and therefore from (3) and (4)

$$|\det A|^{2/n} \leq n\mu^2.$$

Part (b) follows immediately from⁻(3).

In proving Theorem 3.7 (a) we used the familiar *arithmetic-geometric mean inequality*. To be explicit, this result compares the product of n positive numbers with their sum. Thus, if $a_1, \ldots, a_n$ are any n positive numbers then

$$\left(\prod_{i=1}^{n} a_i\right)^{1/n} \leq \frac{1}{n}\left(\sum_{i=1}^{n} a_i\right).$$

Equality holds in the preceding inequality if and only if $a_1 = a_2 = \cdots = a_n$.

Theorem 3.8 *The rank of an n-square normal matrix A is the size of the largest nonzero principal subdeterminant of A.*

Proof. We know from Theorems 3.1 and 3.4 that there exists an n-square unitary matrix U such that $A = U^*DU$ where

$$D = \text{diag }(\lambda_1, \ldots, \lambda_k, 0, \ldots, 0),$$

$\lambda_i \neq 0, i = 1, \ldots, k,$ and $\rho(A) = \rho(D) = k$. Then from Theorem 6.1, Chap. 2, we have, for $\alpha \in Q_{k,n}$,

$$\det (A[\alpha|\alpha]) = \sum_{\omega \in Q_{k,n}} \det (U^*[\alpha|\omega]) \det ((DU)[\omega|\alpha])$$

$$= \sum_{\omega, \gamma \in Q_{k,n}} \det (U^*[\alpha|\omega]) \det (D[\omega|\gamma]) \det (U[\gamma|\alpha]).$$

But unless $\omega = \gamma = (1, \ldots, k)$, $\det (D[\omega|\gamma]) = 0$. Also

$$\det (D[1, \ldots, k|1, \ldots, k]) = \prod_{i=1}^{k} \lambda_i.$$

Hence

$$\det (A[\alpha|\alpha]) = \det (U^*[\alpha|1, \ldots, k]) \det (U[1, \ldots, k|\alpha]) \prod_{i=1}^{k} \lambda_i.$$

$$(5)$$

Now using formula (15), Sec. 2.6, we conclude that

$$\det (U^*[\alpha|1, \ldots, k]) = \overline{\det (U[1, \ldots, k|\alpha])}.$$

Hence (5) becomes

$$\det (A[\alpha|\alpha]) = |\det (U[\alpha|1, \ldots, k])|^2 \prod_{i=1}^{k} \lambda_i.$$

By Theorem 6.7, Chap. 2, we know that there exists $\alpha^0 \in Q_{k,n}$ such that $\det (U[\alpha^0|1, \ldots, k]) \neq 0$. Hence $\det (A[\alpha^0|\alpha^0]) \neq 0$. Since $\rho(A) = k$, no determinant of a $(k + 1)$-square submatrix of A can be nonzero (see Theorem 6.6, Chap. 2).

Definition 3.1 (*Simultaneous triangularization*) *If S is a nonsingular matrix for which $S^{-1}AS$ and $S^{-1}BS$ are both diagonal (triangular), then A and B are said to be* **simultaneously diagonalizable** (**triangularizable**).

Clearly if two matrices are simultaneously diagonalizable then they commute. We shall show in Theorem 3.11 that for normal matrices the converse also holds.

Theorem 3.9 *If $A, B \in M_n(R)$, and $AB = BA$, then A and B have a characteristic vector in common.*

Proof. Let λ_1 be a characteristic root of A and let $x_1, \ldots, x_k$ be a basis for the null space of $\lambda_1 I_n - A$. Then any nonzero linear combination of $x_1, \ldots, x_k$ is a characteristic vector corresponding to λ_1 and conversely any characteristic vector of A corresponding to λ_1 is in $\langle x_1, \ldots, x_k \rangle$.

Now, if for some i, $Bx_i = 0$, then x_i is a characteristic vector of B corresponding to the characteristic root 0 and the theorem is proved. Suppose that $Bx_i \neq 0$, $i = 1, \ldots, k$. Then

$$\begin{aligned} A(Bx_i) &= BAx_i \\ &= B(\lambda_1 x_i) \\ &= \lambda_1(Bx_i) \end{aligned}$$

and thus

$$Bx_i \in \langle x_1, \ldots, x_k \rangle, \quad i = 1, \ldots, k. \tag{6}$$

Let $X \in M_{n,k}(R)$ be the matrix whose jth column is $x_j, j = 1, \ldots, k$.

Then, by Theorem 3.3(c), Chap. 2, the system of equations

$$Xy_i = Bx_i, \quad i = 1, \ldots, k, \tag{7}$$

where $y_i \in V_k(R)$, has a solution for y_i, $i = 1, \ldots, k$. If $Y \in M_k(R)$ is the matrix whose ith column is y_i, $i = 1, \ldots, k$, the relation (7) can be written compactly

$$XY = BX. \tag{8}$$

Let μ_1 be any characteristic root of Y and $z = (\alpha_1, \ldots, \alpha_k)$ a characteristic vector of Y corresponding to μ_1. Then, by (8),

$$B(Xz) = XYz$$

and, since $Yz = \mu_1 z$,

$$B(Xz) = \mu_1(Xz).$$

Now,

$$Xz = \sum_{t=1}^{k} \alpha_t x_t \neq 0$$

because $z = (\alpha_1, \ldots, \alpha_k) \neq 0$, and $x_1, \ldots, x_k$ are linearly independent. Thus Xz is a characteristic vector of B. But

$$Xz = \sum_{t=1}^{k} \alpha_t x_t \in \langle x_1, \ldots, x_k \rangle$$

and therefore Xz is a characteristic vector of A.

It is instructive to illustrate Theorem 3.9 with an example. Let

$$A = \begin{bmatrix} 1 & 0 & -1 & 0 \\ 0 & 1 & 0 & -1 \\ 1 & 0 & -1 & 0 \\ 0 & 1 & 0 & -1 \end{bmatrix} \quad \text{and} \quad B = \begin{bmatrix} 4 & -1 & -1 & 0 \\ -1 & 4 & 0 & -1 \\ 1 & 0 & 2 & -1 \\ 0 & 1 & -1 & 2 \end{bmatrix}.$$

Then

$$AB = \begin{bmatrix} 3 & -1 & -3 & 1 \\ -1 & 3 & 1 & -3 \\ 3 & -1 & -3 & 1 \\ -1 & 3 & 1 & -3 \end{bmatrix} = BA.$$

The problem is to find a characteristic vector common to A and B. The matrix A is clearly singular and therefore $\lambda_1 = 0$ is a characteristic root. Since the rank of $\lambda_1 I_4 - A$ is 2, the dimension of its null space is $4 - 2 = 2$. It is easy to see that the vectors $x_1 = (1,0,1,0)$ and $x_2 = (0,1,0,1)$ form a basis for this null space. Now, we compute

$$Bx_1 = 3x_1 - x_2 \quad \text{and} \quad Bx_2 = -x_1 + 3x_2.$$

Let

$$X = \begin{bmatrix} 1 & 0 \\ 0 & 1 \\ 1 & 0 \\ 0 & 1 \end{bmatrix} \quad \text{and} \quad Y = \begin{bmatrix} 3 & -1 \\ -1 & 3 \end{bmatrix}.$$

Then, as in (8),

$$XY = BX.$$

Clearly $\mu_1 = 4$ is a characteristic root of Y and $z = (1,-1)$ is a corresponding characteristic vector Then $Xz = (1,-1,1,-1)$ is a characteristic vector common to A and B.

Theorem 3.10   *If $A,B \in M_n(R)$, and $AB = BA$, then A and B can be simultaneously unitarily triangularized.*

Proof.   We use induction on n. By Theorem 3.9, there exists $x_1 \in V_n(R)$, of unit length, such that

$$Ax_1 = \lambda_1 x_1 \quad \text{and} \quad Bx_1 = \mu_1 x_1.$$

We proceed as in the proof of Theorem 2.6. We construct a unitary matrix S whose first column is x_1. Then

$$S^*AS = \left[\begin{array}{c|c} \lambda_1 & z \\ \hline 0 & A_1 \end{array} \right] \quad \text{and} \quad S^*BS = \left[\begin{array}{c|c} \mu_1 & w \\ \hline 0 & B_1 \end{array} \right]$$

where A_1 and B_1 are in $M_{n-1}(R)$. Now, $(S^*AS)(S^*BS) = S^*ABS = S^*BAS = (S^*BS)(S^*AS)$, and by block multiplication we have

$$\begin{bmatrix} \lambda_1\mu_1 & \lambda_1 w + zB_1 \\ 0 & A_1B_1 \end{bmatrix} = \begin{bmatrix} \lambda_1\mu_1 & \mu_1 z + wA_1 \\ 0 & B_1A_1 \end{bmatrix}$$

and therefore $A_1B_1 = B_1A_1$. The conclusion of the proof is analogous to that of Theorem 2.6 and is left as an exercise for the student.

We are now able to prove the result about commuting normal matrices that we announced earlier.

Theorem 3.11 *If A and B are n-square normal matrices, then $AB = BA$ if and only if there exists an n-square unitary matrix U such that U^*AU and U^*BU are both diagonal.*

Proof. Suppose first that $AB = BA$. Then according to Theorem 3.10 there exists a unitary matrix U such that U^*AU and U^*BU are both triangular. But as we saw in Theorem 3.1, a normal triangular matrix must in fact be diagonal. Hence U^*AU and U^*BU are diagonal. On the other hand, if $U^*AU = D_1$ and $U^*BU = D_2$ are diagonal matrices, then

$$\begin{aligned} AB &= (UD_1U^*)(UD_2U^*) = UD_1D_2U^* \\ &= UD_2D_1U^* = (UD_2U^*)(UD_1U^*) \\ &= BA. \end{aligned}$$

This completes the proof.

Quiz

Answer true or false:

1. The matrix

$$\begin{bmatrix} 1+i & i \\ i & 1-i \end{bmatrix}$$

is normal.

2. If A and B are normal n-square matrices with the same characteristic vectors, then A and B commute.

3. Any 2-square symmetric (real) orthogonal matrix is a scalar multiple of the identity matrix or of the matrix diag $(1, -1)$.

4. A normal matrix is nilpotent if and only if it is the zero matrix.

5. If A is both hermitian and skew-hermitian, then $A = 0$.
6. A product of normal matrices is normal.
7. An n-square matrix A is normal if and only if $I_n - A$ is normal.
8. An n-square matrix A is skew-hermitian if and only if $I_n - A$ is skew-hermitian.
9. If A is normal and $B^{-1}AB$ is diagonal, then B is unitary.
10. If H and K are hermitian n-square matrices, then the matrix $H + iK$ is normal.

Exercises

1. Show that any 2-square real orthogonal matrix A is of the form,

$$A = \begin{bmatrix} \cos\theta & \sin\theta \\ -\sin\theta & \cos\theta \end{bmatrix}, \quad \text{or} \quad A = \begin{bmatrix} \cos\theta & \sin\theta \\ \sin\theta & -\cos\theta \end{bmatrix}, \quad 0 \le \theta \le 2\pi.$$

2. If U is a unitary n-square matrix show that $|\operatorname{tr} U| \le n$ with equality if and only if $U = cI_n$ for some scalar c of modulus 1.
3. Prove that if A is normal then A^k is normal for any positive integer k.
4. Let H be a real symmetric 2-square matrix with characteristic roots λ_1 and λ_2. Show that

$$\det(H) \le \operatorname{per}(H) \le \tfrac{1}{2}(\lambda_1{}^2 + \lambda_2{}^2).$$

5. Discuss the cases of equality in Exercise 4.
6. Show that if A and B are 2-square complex matrices with the same characteristic vectors, then A and B commute.
7. Given a normal matrix A, construct a normal matrix B such that $B^2 = A$.
8. Complete the proof of Theorem 3.10.
9. Show that the first part of Theorem 2.6 is an immediate consequence of Theorem 3.10.
10. Show by a counterexample that the converse of Theorem 3.10 is false.

3.4 Hermitian Matrices

In the first theorem in this section we collect together some previously established results on hermitian and skew-hermitian matrices. As usual, the field R will be the complex numbers unless otherwise stated.

Theorem 4.1 (a) *Any square complex matrix A is a sum of a hermitian matrix H and a skew-hermitian matrix S:*

$$A = H + S.$$

Moreover, H and S are uniquely determined by A.

(b)   *All the characteristic roots of a hermitian matrix are real.*

(c)   *All the characteristic roots of a skew-hermitian matrix are pure imaginary.*

(d)   *A hermitian (skew-hermitian) matrix has an orthonormal set of characteristic vectors.*

Proof.   (a)   We set (see the proof of Theorem 4.5, Chap. 1)

$$H = \frac{(A + A^*)}{2} \quad \text{and} \quad S = \frac{(A - A^*)}{2},$$

so that $A = H + S$. In order to prove the uniqueness of H and S, suppose that

$$A = H_1 + S_1$$

where

$$H_1{}^* = H_1 \quad \text{and} \quad S_1{}^* = -S_1.$$

Then

$$A^* = H_1 - S_1$$

and

$$H_1 = \frac{(A + A^*)}{2} = H, \quad S_1 = \frac{(A - A^*)}{2} = S.$$

For parts (b), (c) and (d) see Theorems 3.5 and 3.3.

Theorem 4.1(b), (d) can also be expressed as follows.

Theorem 4.2   *For every hermitian linear transformation T of a unitary space V into itself there exists an orthonormal basis G of V such that $[T]_G{}^G$ is a real diagonal matrix.*

The sum of two hermitian matrices is clearly hermitian. However, the product of two hermitian matrices, in general, is not hermitian. For example, both $A = \begin{bmatrix} 2 & 0 \\ 0 & 1 \end{bmatrix}$ and $B = \begin{bmatrix} 2 & 1 \\ 1 & 1 \end{bmatrix}$ are hermitian but neither $AB = \begin{bmatrix} 4 & 2 \\ 1 & 1 \end{bmatrix}$ nor $BA = \begin{bmatrix} 4 & 1 \\ 2 & 1 \end{bmatrix}$ is hermitian.

Theorem 4.3 *The product of two n-square hermitian matrices is hermitian if and only if the matrices commute.*

Proof. Let A and B be the two hermitian matrices. Suppose $AB = BA$. Then

$$(AB)^* = (BA)^* = A^*B^* = AB$$

and AB is hermitian. Conversely, suppose that $AB = (AB)^*$. Then

$$AB = (AB)^* = B^*A^* = BA.$$

In other words, A and B commute.

Note that in the proof we only made use of the formulas $(AB)^* = B^*A^*$, $A = A^*$, $B = B^*$ and thus there is an analogous theorem for linear transformations.

Theorem 4.4 *Let A be an n-square hermitian matrix and let T_A be the linear transformation on $V_n(R)$ defined by*

$$T_A x = Ax$$

for all $x \in V_n(R)$. Then T_A is hermitian with respect to the standard inner product in $V_n(R)$ and has the same characteristic roots as A.

Proof. Let x and y be any vectors in $V_n(R)$. Then using the standard inner product in $V_n(R)$ we compute

$$\begin{aligned}(T_A x, y) &= (Ax, y)\\ &= (x, A^*y)\\ &= (x, Ay)\\ &= (x, T_A y).\end{aligned}$$

Thus T_A is hermitian by Theorem 4.3, Chap. 1. Now let λ_i be a characteristic root of A and $u_i \in V_n(R)$ be a corresponding characteristic vector. Then

$$A u_i = \lambda_i u_i.$$

Therefore

$$T_A u_i = \lambda_i u_i$$

and λ_i is a characteristic root of T_A.

Theorem 4.5   Let $T \in L(V,V)$ be a normal transformation. Let x_1, . . . , x_n be an orthonormal basis of V. Then the matrix

$$A = ((Tx_i,x_j))$$

is normal and has the same characteristic roots as T. The matrix A is hermitian (skew-hermitian, unitary) if and only if T has the same property.

Proof.   Let $X = \{x_1, . . . , x_n\}$. Then, by Theorem 4.2, Chap. 1,

$$[T]_X{}^X = A^T.$$

Hence, by Theorem 4.4, Chap. 1, the matrix A^T is normal. It follows that A is also normal. Also, by the same theorem, A is hermitian (skew-hermitian, unitary) if and only if T is. We also know (see Sec. 3.1) that the matrix A has the same characteristic roots as T.

Many of the properties of hermitian matrices have analogues for skew-hermitian matrices. This is immediate because A is hermitian if and only if iA is skew-hermitian. However, if we restrict ourselves to real matrices then the real skew-hermitian, i.e., the skew-symmetric matrices, have certain properties which are peculiarly their own. One of them is the following.

Theorem 4.6   A nonzero (real) skew-symmetric n-square matrix S is orthogonally similar to a real matrix of the form

$$\begin{bmatrix} 0 & a_1 \\ -a_1 & 0 \end{bmatrix} + \cdots + \begin{bmatrix} 0 & a_m \\ -a_m & 0 \end{bmatrix} + 0_{n-2m,\,n-2m} \qquad (1)$$

where $a_i \neq 0$, $i = 1, . . . , m$.

Proof.   We prove the theorem by induction on n. If all characteristic roots of S are 0 then S is zero because, by Theorem 3.1, S is similar to a zero matrix. Otherwise, by Theorem 4.1(c), S has a nonzero pure imaginary characteristic root ia_1. Let the corresponding characteristic vector be $x_1 + iy_1$, where x_1 and y_1 are real n-tuples.

First observe that $S(x_1 + iy_1) = ia_1(x_1 + iy_1)$ implies

$$Sx_1 = -a_1 y_1 \qquad \text{and} \qquad Sy_1 = a_1 x_1.$$

Therefore, using the standard inner product in $V_n(R)$, R the real number field, we have

$$\begin{aligned}
a_1(x_1,x_1) &= (Sy_1,x_1) \\
&= (y_1, S^T x_1) \\
&= -(y_1, S x_1) \\
&= -(S x_1, y_1) \\
&= a_1(y_1,y_1)
\end{aligned}$$

and thus

$$\|x_1\| = \|y_1\| \neq 0.$$

Moreover,

$$\begin{aligned}
(x_1,y_1) &= \frac{1}{a_1}(Sy_1,y_1) \\
&= \frac{1}{a_1}(y_1, S^T y_1) \\
&= \frac{1}{a_1}(y_1, -Sy_1) \\
&= -\frac{1}{a_1}(Sy_1,y_1) \\
&= -(x_1,y_1)
\end{aligned}$$

and therefore $(x_1,y_1) = 0$. Now, let $u_1 = x_1/\|x_1\|$, $u_2 = y_1/\|y_1\|$ and complete u_1, u_2 to an orthonormal set of real vectors $u_1, u_2, u_3, \ldots, u_n$. Let U be the n-square orthogonal matrix such that

$$U^{(j)} = u_j, \qquad j = 1, \ldots, n.$$

Then we assert that

$$U^T S U = \left[\begin{array}{cc:c} 0 & a_1 & \\ -a_1 & 0 & \raisebox{1.5ex}{0} \\ \hdashline & 0 & S_1 \end{array}\right]$$

where S_1 is an $(n - 2)$-square skew symmetric real matrix. For,

$$
\begin{aligned}
(U^T S U)_{1,j} &= (S u_j, u_1) \\
&= (u_j, S^T u_1) \\
&= -(u_j, S u_1) \\
&= -(u_j, -a_1 u_2) \\
&= a_1(u_j, u_2) \\
&= a_1 \delta_{j2}.
\end{aligned}
$$

Similarly,

$$
(U^T S U)_{2,j} = -a_1 \delta_{1j}.
$$

Moreover,

$$
(U^T S U)^T = U^T S^T U = -U^T S U
$$

so that $U^T S U$ is skew-symmetric and thus the other parts of our assertion are also justified. Now, by the induction hypothesis, there exists a real orthogonal $(n - 2)$-square matrix W_1 such that

$$
W_1^T S_1 W_1 = \begin{bmatrix} 0 & a_2 \\ -a_2 & 0 \end{bmatrix} + \cdots + \begin{bmatrix} 0 & a_m \\ -a_m & 0 \end{bmatrix} + 0_{n-2m, n-2m}.
$$

Let $W = I_2 + W_1$. Then $V = UW$ is real orthogonal and

$$
V^T S V = \left[\begin{array}{cc:c} 0 & a_1 & \\ -a_1 & 0 & 0 \\ \hdashline & 0 & W_1^T S_1 W \end{array} \right]
$$

$$
= \begin{bmatrix} 0 & a_1 \\ -a_1 & 0 \end{bmatrix} + \cdots + \begin{bmatrix} 0 & a_m \\ -a_m & 0 \end{bmatrix} + 0_{n-2m, n-2m},
$$

and the induction is complete.

We saw in Theorem 4.1 (a) that any square complex matrix A is a sum of a hermitian and a skew-hermitian matrix, or (since K is hermitian if and only if $S = iK$ is skew-hermitian),

$$
A = H + iK,
$$

where H and K are uniquely determined hermitian matrices. For 1-square matrices this statement reduces to the decomposition of a complex number into real and imaginary parts. We shall introduce in Theorem 4.13 a polar decomposition of a complex matrix analogous to the decomposition of a complex number into a product of a non-negative real number and a complex number of modulus 1.

Definition 4.1 (*Positive-definite*) *A hermitian matrix, or a hermitian transformation, is called* **positive definite** *if all its characteristic roots are positive. It is said to be* **positive semidefinite** *if all its characteristic roots are nonnegative. A positive semidefinite hermitian matrix or transformation is also called* **nonnegative hermitian**.

Definition 4.2 (*Gram matrix*) *Let* $x_1, \ldots, x_n$ *be vectors in a unitary space* V. *The n-square hermitian matrix whose* (i,j) *entry is* (x_i, x_j) *is called the* **Gram matrix** *based on* $x_1, \ldots, x_n$.

Theorem 4.7 *Let* V *be any unitary space of dimension not less than* n. *A necessary and sufficient condition for an n-square matrix to be hermitian positive semidefinite is that it be a Gram matrix based on* n *vectors in* V.

Proof.   Let $H = (h_{ij})$ be a positive semidefinite n-square matrix. Then, by Theorem 3.1, there exists a unitary matrix U such that

$$U^*HU = \text{diag } (\lambda_1, \ldots, \lambda_n)$$

where $\lambda_j \geq 0$, $j = 1, \ldots, n$. Let $D = \text{diag } (\sqrt{\lambda_1}, \ldots, \sqrt{\lambda_n})$ and let

$$A = UDU^*.$$

Then

$$AA^* = UDU^*UDU^*$$
$$= UD^2U^*$$
$$= H.$$

Now let $v_1, \ldots, v_n$ be an orthonormal set of vectors in V. Set

$$x_i = \sum_{s=1}^{n} a_{is}v_s.$$

Then

$$(x_i, x_j) = \Big(\sum_{s=1}^{n} a_{is}v_s, \sum_{t=1}^{n} a_{jt}v_t \Big)$$

$$= \sum_{s,t=1}^{n} a_{is}\bar{a}_{jt}(v_s, v_t)$$

$$= \sum_{s=1}^{n} a_{is}\bar{a}_{js}$$

$$= (AA^*)_{ij}$$

$$= h_{ij}.$$

Conversely, suppose that $h_{ij} = (x_i, x_j)$. Let $v_1, \ldots, v_n$ be an orthonormal set in V, and suppose that

$$x_i = \sum_{t=1}^{n} a_{it}v_t, \quad i = 1, \ldots, n.$$

Set

$$A = (a_{ij}) \in M_n(R).$$

Then

$$h_{ij} = (x_i, x_j) = \Big(\sum_{s=1}^{n} a_{is}v_s, \sum_{t=1}^{n} a_{jt}v_t \Big)$$

$$= \sum_{s,t=1}^{n} a_{is}\bar{a}_{jt}(v_s, v_t)$$

$$= \sum_{s=1}^{n} a_{is}\bar{a}_{js}$$

$$= (AA^*)_{ij}.$$

Thus

$$H = AA^*.$$

Since $AA^* = (AA^*)^*$, H is hermitian and there exists a unitary matrix

U such that U^*AA^*U is diagonal, i.e.,

$$BB^* = \text{diag}\,(\mu_1, \ldots, \mu_n)$$

where $B = (b_{ij}) = U^*A$ and $\mu_1, \ldots, \mu_n$ are the characteristic roots of H. Then

$$
\begin{aligned}
\mu_i &= \sum_{t=1}^{n} B_{it}(B^*)_{ti} \\
&= \sum_{t=1}^{n} b_{it}\bar{b}_{it} \\
&= \sum_{t=1}^{n} |b_{it}|^2 \geq 0, \quad i = 1, \ldots, n,
\end{aligned}
$$

and H is positive semidefinite.

In Theorem 4.7 we can choose the space V to be $V_n(R)$. We can then state the result in a somewhat more special form.

Theorem 4.8 *An n-square matrix H is positive semidefinite if and only if there exists $A \in M_n(R)$ such that $H = AA^*$.*

Proof. Assume H is positive semidefinite. Let $A_{(i)}$ be the vector $x_i \in V_n(R)$ in Theorem 4.7. Then $H = ((x_i, x_j)) = AA^*$. Conversely, if $H = AA^*$ then $h_{ij} = (A_{(i)}, A_{(j)})$, $i,j = 1, \ldots, n$.

A positive semidefinite hermitian matrix is positive definite if and only if it is nonsingular (see Theorem 3.4).

Theorem 4.9 *An n-square hermitian matrix H is positive definite if and only if it is a Gram matrix based on a set of n linearly independent vectors. Alternatively, H is positive definite if and only if there exists a nonsingular matrix A such that $H = AA^*$.*

Proof. We know (see Theorem 6.8 and Exercise 2, Sec. 2.6) that $\rho(A) = \rho(AA^*)$ and this completes this proof.

In the course of the proof of Theorem 4.7 we constructed a positive semidefinite matrix A such that $H = A^2$. This matrix plays a similar role to that of a square root of a nonnegative number.

Definition 4.3 (**Square root**) *If H and A are positive semidefinite hermitian matrices and $H = A^2$, then A is called the **square root** of H and is denoted by $H^{1/2}$ or $\sqrt{H}$.*

We now justify the use of the definite article "the" before "square root" in the above definition.

Theorem 4.10 *Any positive semidefinite hermitian matrix has a unique square root.*

Proof. Let H be the positive semidefinite matrix and let U be unitary so that

$$U^*HU = \text{diag }(\lambda_1, \ldots, \lambda_n)$$

where $\lambda_j \geq 0$, $j = 1, \ldots, n$, and $\{U^{(1)}, \ldots, U^{(n)}\}$ is a set of orthonormal characteristic vectors of H. Set

$$A = U \text{ diag }(\sqrt{\lambda_1}, \ldots, \sqrt{\lambda_n})U^*.$$

Then A is positive semidefinite hermitian and

$$A^2 = H,$$

i.e.,

$$A = H^{1/2}.$$

Now, suppose that A and B are positive semidefinite matrices and

$$H = A^2 = B^2.$$

Suppose that $\lambda_1 \geq \cdots \geq \lambda_r > 0$ and $\lambda_{r+1} = \cdots = \lambda_n = 0$, where $r = \rho(H)$. Since $U^{(j)}$, $j = 1, \ldots, n$, are the characteristic vectors of H, we have

$$A^2U^{(j)} = \lambda_j U^{(j)}, \quad j = 1, \ldots, n,$$

or

$$(\lambda_j I_n - A^2)U^{(j)} = (\sqrt{\lambda_j} I_n + A)(\sqrt{\lambda_j} I_n - A)U^{(j)} = 0. \quad (2)$$

Now the negative numbers $-\sqrt{\lambda_1}, \ldots, -\sqrt{\lambda_r}$ cannot be characteristic roots of A and therefore $\sqrt{\lambda_j} I_n + A$ is nonsingular for $j = 1, \ldots, r$. Thus (2) becomes

$$(\sqrt{\lambda_j} I_n - A) U^{(j)} = 0, \quad j = 1, \ldots, r,$$

and $U^{(1)}, \ldots, U^{(r)}$ are characteristic vectors of A corresponding to $\sqrt{\lambda_1}, \ldots, \sqrt{\lambda_r}$ respectively. Moreover, using the standard inner product in $V_n(R)$, we have

$$\|A U^{(j)}\|^2 = (A U^{(j)}, A U^{(j)}) = (A^* A U^{(j)}, U^{(j)}) = (A^2 U^{(j)}, U^{(j)})$$
$$= (H U^{(j)}, U^{(j)}).$$

But $H U^{(j)} = 0$, $j = r + 1, \ldots, n$, and hence $A U^{(j)} = 0$, $j = r + 1, \ldots, n$. It follows that

$$U^* A U = \text{diag } (\sqrt{\lambda_1}, \ldots, \sqrt{\lambda_r}, 0, \ldots, 0);$$

similarly,

$$U^* B U = \text{diag } (\sqrt{\lambda_1}, \ldots, \sqrt{\lambda_r}, 0, \ldots, 0).$$

Hence

$$A = B.$$

We now state and prove two important theorems that give necessary and sufficient conditions for a matrix to be positive semidefinite (definite).

Theorem 4.11   *A hermitian matrix is positive semidefinite (definite) if and only if the determinants of all its principal submatrices are nonnegative (positive).*

Proof.   Suppose H is positive semidefinite. Thus H is a Gram matrix based on a set of vectors $x_1, \ldots, x_n$. Then for any sequence $\omega = (\omega_1, \ldots, \omega_k) \in Q_{k,n}$, $1 \le k \le n$, the principal submatrix $H[\omega|\omega]$ is a Gram matrix based on the vectors $x_{\omega_1}, \ldots, x_{\omega_k}$ and is therefore positive semidefinite. Hence det $(H[\omega|\omega]) \ge 0$ for all $\omega \in Q_{k,n}$, $k = 1, \ldots, n$. If H also happens to be nonsingular and therefore positive definite, then by Theorem 4.9, $x_1, \ldots, x_n$ are linearly independent. Hence, a fortiori, $x_{\omega_1}, \ldots, x_{\omega_k}$ are linearly independent and $\det(H[\omega|\omega]) > 0$ for all $\omega \in Q_{k,n}$, $k = 1, \ldots, n$.

Conversely, suppose that the determinants of all principal submatrices of H are nonnegative. Let $\lambda_1 \geq \cdots \geq \lambda_n$ be the characteristic roots of H. By Theorem 1.4,

$$E_r(H) \geq 0, \quad r = 1, \ldots, n. \tag{3}$$

We have

$$\prod_{i=1}^{n} (\lambda + \lambda_i) = \lambda^n + \sum_{r=1}^{n} E_r(H)\lambda^{n-r}. \tag{4}$$

Now suppose that $\lambda_n < 0$. Set $\lambda = -\lambda_n > 0$ in (4). Then the left-hand side is 0 while the right-hand side is not less than $(-\lambda_n)^n$, a positive number. Thus the assumption $\lambda_n < 0$ leads to a contradiction.

If, instead of (3), we have

$$E_r(H) > 0, \quad r = 1, \ldots, n,$$

then in particular

$$E_n(H) = \prod_{j=1}^{n} \lambda_j > 0$$

and the positive semidefinite matrix H has no zero characteristic roots and is therefore positive definite.

In the next section we shall refine the conditions of Theorem 4.11. In the remainder of this section we use the standard inner product in $V_n(R)$, R the complex-number field.

Theorem 4.12 *A hermitian n-square matrix H is positive definite (semidefinite) if and only if $(Hx,x) > 0$ $[(Hx,x) \geq 0]$ for all nonzero $x \in V_n(R)$.*

Proof. Let U be a unitary matrix such that

$$U^*HU = \text{diag}(\lambda_1, \ldots, \lambda_n)$$

where $\lambda_1 \geq \cdots \geq \lambda_n$ are the characteristic roots of H. Let x be any nonzero vector in $V_n(R)$, and let $y = (y_1 \cdots y_n) = U^*x$. Then

$$
\begin{aligned}
(Hx,x) &= (HUy,Uy) \\
&= (U^*HUy,y) \\
&= \sum_{i=1}^{n} \lambda_i y_i \bar{y}_i \\
&= \sum_{i=1}^{n} \lambda_i |y_i|^2.
\end{aligned}
$$

Since $x \neq 0$ and therefore $y \neq 0$, we can conclude that $(Hx,x) > 0$ for all nonzero x $[(Hx,x) \geq 0$ for all $x]$ if all the characteristic roots of H are positive (nonnegative), i.e., if H is positive definite (semidefinite).

To prove the converse, suppose that H is not positive definite, i.e., $\lambda_n \leq 0$. Let x_n be a corresponding characteristic vector of length 1. Then $Hx_n = \lambda_n x_n$ and hence $(Hx_n,x_n) = \lambda_n(x_n,x_n) = \lambda_n \leq 0$ and (Hx,x) cannot be positive for all nonzero x. Similarly, if H is not positive semidefinite, i.e., $\lambda_n < 0$, then $(Hx,x) < 0$ for $x = x_n$.

Our next result is the so-called *polar factorization theorem*.

Theorem 4.13 *Let A be any complex n-square matrix. Then there exist positive semidefinite hermitian matrices H and K and unitary matrices U and V such that*

$$
A = UH = KV. \tag{5}
$$

The matrices H and K are uniquely determined. If A is nonsingular, then U and V are also uniquely determined. A is normal if and only if $A = UH = HU = KV = VK$.

Proof. Let $\alpha_1{}^2 \geq \cdots \geq \alpha_n{}^2$ be the characteristic roots of the positive semidefinitive matrix A^*A (see Theorem 4.8). Let $\alpha_i > 0$, $i = 1, \ldots, r$, and $\alpha_i = 0$, $i = r+1, \ldots, n$. By Theorem 4.1 (d), there exists a set of orthonormal characteristic vectors of A^*A. Let

$$
A^*Ax_i = \alpha_i{}^2 x_i, \qquad i = 1, \ldots, n, \quad (x_i,x_j) = \delta_{ij}.
$$

Let

$$
z_j = Ax_j/\alpha_j, \qquad j = 1, \ldots, r.
$$

Then, for $1 \leq i, j \leq k$,

$$
\begin{aligned}
(z_i, z_j) &= \frac{1}{\alpha_i \alpha_j} (Ax_i, Ax_j) \\
&= \frac{1}{\alpha_i \alpha_j} (A^*Ax_i, x_j) \\
&= \frac{1}{\alpha_i \alpha_j} (\delta_{ij}\alpha_i{}^2)
\end{aligned}
$$

and thus $z_1, \ldots, z_r$ are orthonormal. Let X and Z be unitary n-square matrices such that

$$X^{(j)} = x_j, \quad j = 1, \ldots, n,$$

and

$$Z^{(j)} = z_j, \quad j = 1, \ldots, r.$$

Then

$$Z^{(j)} = AX^{(j)}/\alpha_j, \quad j = 1, \ldots, r,$$

or, if $D = \text{diag}(\alpha_1, \ldots, \alpha_n)$,

$$ZD = AX.$$

Now, let $U = ZX^*$ and $H = XDX^*$. Thus U is unitary and H is positive semidefinite hermitian. Also,

$$UH = ZX^*XDX^* = ZDX^* = A.$$

Similarly,

$$A^* = V^*K$$

where V^* is unitary and K is positive semidefinite hermitian. Thus

$$A = KV.$$

Now, suppose that $A = U_1H_1 = U_2H_2$ where U_1, U_2 are unitary and

H_1, H_2 are positive semidefinite hermitian. Then

$$A^*A = (U_1H_1)^*U_1H_1 = H_1^*H_1 = H_1^2 = (U_2H_2)^*U_2H_2 = H_2^2.$$

Thus, by Theorem 4.10,

$$H_1 = H_2 = (A^*A)^{1/2}.$$

Similarly, if $A = K_1V_1 = K_2V_2$, where K_1,K_2 are positive semidefinite hermitian and V_1,V_2 are unitary, then

$$K_1 = K_2 = (AA^*)^{1/2}.$$

If $A = U_1H = U_2H = KV_1 = KV_2$, where U_1,U_2,V_1,V_2 are unitary and H,K are positive definite hermitian, then

$$U_1 = U_2 = AH^{-1}$$

and

$$V_1 = V_2 = K^{-1}A.$$

Lastly, if $UH = HU$ then

$$AA^* = UH^2U^* = HUHU^* = H^2UU^* = H^2 = A^*A$$

and A is normal. Conversely, if A is normal then

$$AA^* = UH^2U^* = A^*A = H^2$$

or

$$(UHU^*)^2 = H^2.$$

Therefore, by Theorem 4.10,

$$UHU^* = H,$$

or

$$UH = HU.$$

We leave the proof of the following theorem as an exercise for the student.

Theorem 4.14 *Let A be any real n-square matrix. Then there exist unique positive semidefinite symmetric matrices H and K and (real) orthogonal matrices U and V such that*

$$A = UH = KV.$$

If A is nonsingular, then U and V are uniquely determined. A is normal if and only if $A = UH = HU$.

Definition 4.4 *(Singular values)* *If $A = UH$ where U is unitary and H is positive semidefinite hermitian then the characteristic roots of H are called the **singular values** of A.*

Quiz

Answer true or false:

1. If all the characteristic roots of an n-square real matrix A are positive and all the entries in $x \in V_n(R)$ are positive then $(Ax,x) > 0$.
2. If A is nonsingular hermitian and both AB and BA are hermitian, then B is hermitian.
3. If all the roots of a skew-hermitian matrix S are equal then S is a multiple of the identity matrix.
4. Any $(2m + 1)$-square real skew-hermitian matrix is singular.
5. The matrix $\begin{bmatrix} 4 & 4 & 4 \\ 4 & 3 & 2 \\ 4 & 2 & 1 \end{bmatrix}$ is positive semidefinite.
6. If H is a positive definite hermitian n-square matrix then there are 2^n distinct hermitian matrices A such that $A^2 = H$.
7. A matrix A is 0 if and only if all its singular values are 0.
8. If H is hermitian with characteristic roots $\lambda_1 \geq \cdots \geq \lambda_n > 0$, then there exists a unique unitary U such that $U^*HU = \text{diag}(\lambda_1, \ldots, \lambda_n)$.
9. If $(Hx,x) = 0$, where

$$H = \begin{bmatrix} 2 & 1 & 1 \\ 1 & 1 & 1 \\ 1 & 1 & 1 \end{bmatrix}$$

   and $x \in V_3(R)$, then $x = 0$.
10. If $A^*A = B^*B$ is nonsingular, then $A = B$.

Exercises

1. Prove Theorem 4.14.
2. Prove that the rank of a (real) skew-symmetric matrix is an even integer.

3. Show that if $A = KV$ where K is positive semidefinite hermitian and V is unitary then the characteristic roots of K are the singular values of A.

4. Let J be the n-square matrix all of whose entries are 1. Show that $I_n + xJ$ is positive definite if and only if $x > -1/n$.

5. Find the square root of

$$H = \begin{bmatrix} 8 & -2 & -2 \\ -2 & 5 & -4 \\ -2 & -4 & 5 \end{bmatrix}.$$

6. Let

$$A = \begin{bmatrix} 1 & 1 & 1 \\ 1 & -1 & 1 \\ 0 & 0 & 0 \end{bmatrix}.$$

Find unitary matrices U, V and positive semidefinite hermitian matrices H, K such that $A = UH = KV$.

3.5   Matrix Inequalities

The preceding pages contain most of the material usually encountered in a first course in linear algebra. In this final section we display a few samples from parts of linear algebra which are important in applications and yet are still in the forefront of research activity. Throughout this section, R is the field of complex numbers and with the exception of Theorem 5.8, we will be using the standard inner product in $V_n(R)$. We first obtain a simple preliminary result.

Theorem 5.1 *Let* $\lambda_1 \geq \cdots \geq \lambda_n$ *be any real numbers and let* $\sigma_1, \ldots, \sigma_n$ *be nonnegative real numbers satisfying*

$$\sum_{j=1}^{n} \sigma_j = 1.$$

Then

$$\lambda_n \leq \sum_{j=1}^{n} \sigma_j \lambda_j \leq \lambda_1.$$

Proof.   We have

$$\lambda_n \leq \lambda_j \leq \lambda_1, \quad j = 1, \ldots, n,$$

and since $\sigma_j \geq 0$,

$$\sigma_j \lambda_n \leq \sigma_j \lambda_j \leq \sigma_j \lambda_1, \quad j = 1, \ldots, n.$$

Therefore

$$\sum_{j=1}^{n} \sigma_j \lambda_n \leq \sum_{j=1}^{n} \sigma_j \lambda_j \leq \sum_{j=1}^{n} \sigma_j \lambda_1,$$

$$\lambda_n \sum_{j=1}^{n} \sigma_j \leq \sum_{j=1}^{n} \sigma_j \lambda_j \leq \lambda_1 \sum_{j=1}^{n} \sigma_j,$$

$$\lambda_n \leq \sum_{j=1}^{n} \sigma_j \lambda_j \leq \lambda_1.$$

Theorem 5.2   (a)   *If H is a hermitian matrix with characteristic roots $\lambda_1 \geq \cdots \geq \lambda_n$ and $x \in V_n(R)$ is any unit vector then*

$$\lambda_n \leq (Hx,x) \leq \lambda_1.$$

(b)   *If H is the matrix defined in* (a) *and c is any number in the closed interval $[\lambda_n, \lambda_1]$ then there exists a unit vector $y \in V_n(R)$ such that*

$$(Hy,y) = c.$$

Proof. (a)   Let U be a unitary matrix such that $U^*HU = \text{diag}(\lambda_1, \ldots, \lambda_n) = D$, and let $z = (z_1, \ldots, z_n) = U^*x$. Then

$$\|z\|^2 = \sum_{j=1}^{n} |z_j|^2 = \|x\|^2 = 1.$$

We compute that

$$\begin{aligned} (Hx,x) &= (HUz,Uz) \\ &= (U^*HUz,z) \\ &= (Dz,z) \\ &= \sum_{j=1}^{n} \lambda_j |z_j|^2. \end{aligned}$$

Now

$$\sum_{j=1}^{n} |z_j|^2 = 1$$

and thus, by Theorem 5.1,

$$\lambda_n \le \sum_{j=1}^{n} \lambda_j |z_j|^2 \le \lambda_1,$$

or

$$\lambda_n \le (Hx,x) \le \lambda_1.$$

(b)   If $\lambda_1 = \lambda_n$ then, for some unitary matrix U,

$$H = U^* \operatorname{diag}(\lambda_1, \ldots, \lambda_n)U = \lambda_1 U^* U = \lambda_1 I_n$$

and the result holds trivially. Thus assume $\lambda_1 > \lambda_n$ and let

$$\theta = \frac{c - \lambda_n}{\lambda_1 - \lambda_n}.$$

Then $0 \le \theta \le 1$ and

$$c = \theta\lambda_1 + (1 - \theta)\lambda_n.$$

Let $x_1, \ldots, x_n$ be an orthonormal set of characteristic vectors corresponding to $\lambda_1, \ldots, \lambda_n$ and let

$$y = \sqrt{\theta}\, x_1 + \sqrt{(1 - \theta)}\, x_n.$$

Then

$$
\begin{aligned}
(Hy,y) &= (\sqrt{\theta}\, Hx_1 + \sqrt{(1 - \theta)}\, Hx_n, \sqrt{\theta}\, x_1 + \sqrt{(1 - \theta)}\, x_n) \\
&= (\sqrt{\theta}\, \lambda_1 x_1 + \sqrt{(1 - \theta)}\, \lambda_n x_n, \sqrt{\theta}\, x_1 + \sqrt{(1 - \theta)}\, x_n) \\
&= \theta\lambda_1(x_1,x_1) + \sqrt{\theta}\,\sqrt{(1 - \theta)}\,\lambda_1(x_1,x_n) \\
&\qquad + \sqrt{\theta}\,\sqrt{(1 - \theta)}\,\lambda_n(x_n,x_1) + (1 - \theta)\lambda_n(x_n,x_n) \\
&= \theta\lambda_1 + (1 - \theta)\lambda_n \\
&= c.
\end{aligned}
$$

The problem of computing the characteristic roots of a given matrix is usually beyond the capacity of even modern electronic computers. However, it is sometimes important to have available comparisons of the characteristic roots with other more readily computable quantities. In the next few theorems we give several results on the localization of characteristic roots, all obtained in the first half of this century.

Theorem 5.3 *If* $\lambda_1, \ldots, \lambda_n$ *are the characteristic roots of* $A \in M_n(R)$ *and* $\mu_1 \geq \cdots \geq \mu_n$, $\nu_1 \geq \cdots \geq \nu_n$, $\alpha_1 \geq \cdots \geq \alpha_n \geq 0$ *are the characteristic roots of the hermitian matrices* $B = \dfrac{(A + A^*)}{2}$, $C = \dfrac{(A - A^*)}{2i}$ *and* $H = (A^*A)^{\frac{1}{2}}$ *respectively, then*

$$\mu_n \leq \text{Re } (\lambda_t) \leq \mu_1, \qquad t = 1, \ldots, n, \tag{1}$$
$$\nu_n \leq \text{Im } (\lambda_t) \leq \nu_1, \qquad t = 1, \ldots, n, \tag{2}$$
$$\alpha_n \leq |\lambda_t| \leq \alpha_1, \qquad t = 1, \ldots, n. \tag{3}$$

Proof. Let x be a characteristic vector of A of unit length corresponding to λ_t. Then

$$(Ax,x) = \lambda_t(x,x) = \lambda_t$$

and

$$(A^*x,x) = (x,Ax) = \bar{\lambda}_t.$$

Therefore

$$\text{Re } (\lambda_t) = \frac{1}{2}[(Ax,x) + (A^*x,x)] = \left(\frac{A + A^*}{2} x,x\right) = (Bx,x)$$

and

$$\text{Im } (\lambda_t) = \frac{1}{2i}[(Ax,x) - (A^*x,x)] = (Cx,x).$$

Now apply Theorem 5.2 to the hermitian matrices B and C to obtain

$$\mu_n \leq \text{Re } (\lambda_t) = (Bx,x) \leq \mu_1,$$

and

$$\nu_n \leq \text{Im } (\lambda_t) = (Cx,x) \leq \nu_1.$$

To prove (3), note that

$$(A^*Ax,x) = (Ax,Ax) = (\lambda_i x, \lambda_i x) = |\lambda_i|^2$$

and apply Theorem 5.2 to the hermitian matrix A^*A whose characteristic roots are $\alpha_1^2 \geq \cdots \geq \alpha_n^2$ (see Theorem 2.7):

$$\alpha_n^2 \leq (A^*Ax,x) = |\lambda_i|^2 \leq \alpha_1^2.$$

In the next three theorems we require the following special notation. If $A = (a_{ij}) \in M_n(R)$ then

$$P_i = \sum_{\substack{t=1 \\ t \neq i}}^{n} |a_{it}| \quad \text{and} \quad Q_j = \sum_{\substack{t=1 \\ t \neq j}}^{n} |a_{tj}|.$$

Theorem 5.4   *If* $A = (a_{ij}) \in M_n(R)$ *and*

$$|a_{ii}| > P_i, \qquad i = 1, \ldots, n, \tag{4}$$

or

$$|a_{jj}| > Q_j, \qquad j = 1, \ldots, n, \tag{5}$$

then

$$\det (A) \neq 0.$$

Proof.   Suppose (4) holds and $\det (A) = 0$. Let $x = (x_1, \ldots, x_n)$ be a characteristic vector of A corresponding to the characteristic root 0. Let x_k be the largest in absolute value among the numbers $x_1, \ldots, x_n$, i.e.,

$$|x_k| \geq |x_i|, \quad i = 1, \ldots, n.$$

Then

$$A_{(k)}x = \sum_{t=1}^{n} a_{kt}x_t = 0$$

and therefore

$$
\begin{aligned}
|a_{kk}|\,|x_k| &= \left| - \sum_{\substack{t=1 \\ t \neq k}}^{n} a_{kt}x_t \right| \\
&\leq \sum_{\substack{t=1 \\ t \neq k}}^{n} |a_{kt}|\,|x_t| \\
&\leq \sum_{\substack{t=1 \\ t \neq k}}^{n} |a_{kt}|\,|x_k| \\
&= |x_k|P_k.
\end{aligned}
$$

Since $|x_k| > 0$ it follows that

$$|a_{kk}| \leq P_k,$$

contradicting (4).

We prove (5) by applying (4) to the matrix A^T.

The following famous result due to S. A. Geršgorin is almost an immediate consequence of Theorem 5.4.

Theorem 5.5   *The characteristic roots of $A \in M_n(R)$ lie in the closed region of the complex plane consisting of all the disks*

$$|z - a_{ii}| \leq P_i, \qquad i = 1, \ldots, n.$$

They also lie in the closed region consisting of all the disks

$$|z - a_{jj}| \leq Q_j, \qquad j = 1, \ldots, n.$$

Proof.   Let λ_t be a characteristic root of A. Then

$$\det (\lambda_t I_n - A) = 0$$

and, by Theorem 5.4,

$$|\lambda_t - a_{ii}| \leq P_i$$

for at least one i.

The second part of the theorem is proved by applying the first part to the matrix A^T.

The above result has been improved and refined in many ways. The following theorem replacing circular disks in Theorems 5.4 and 5.5 by generally smaller "Cassini ovals" is due to A. Ostrowski and A. Brauer.

Theorem 5.6   (a)   *If $A = (a_{ij}) \in M_n(R)$ and*

$$|a_{ii}|\,|a_{jj}| > P_iP_j, \qquad i,j = 1, \ldots, n, \quad i \neq j, \qquad (6)$$

or

$$|a_{ii}|\,|a_{jj}| > Q_iQ_j, \qquad i,j = 1, \ldots, n, \quad i \neq j, \qquad (7)$$

then

$$\det (A) \neq 0.$$

(b)   *Each characteristic root of A lies in at least one of the $n(n-1)/2$ ovals of Cassini*

$$|z - a_{ii}|\,|z - a_{jj}| \leq P_iP_j, \qquad i,j = 1, \ldots, n, \quad i \neq j, \qquad (8)$$

and in at least one of the $n(n-1)/2$ ovals

$$|z - a_{ii}|\,|z - a_{jj}| \leq Q_iQ_j, \qquad i,j = 1, \ldots, n, \quad i \neq j. \qquad (9)$$

Proof.   (a)   Suppose that the inequalities (6) hold and $\det (A) = 0$. Let $x = (x_1, \ldots, x_n)$ be a characteristic vector of A corresponding to the characteristic root 0. Let k and h be the integers for which

$$|x_k| \geq |x_h| \geq |x_i|, \qquad i = 1, \ldots, k-1, \quad k+1, \ldots, n, \quad k \neq h.$$

Note that $x_h \neq 0$. For if $x_h = 0$, then $x_i = 0$ for $i = 1, \ldots, k-1$, $k+1, \ldots, n$ and $x_k \neq 0$. But this and

$$A_{(k)}x = \sum_{j=1}^{n} a_{kj}x_j = 0$$

imply that $a_{kk}x_k = 0$ and $a_{kk} = 0$ which contradicts (6). Now,

$$A_{(k)}x = A_{(h)}x = 0,$$

or

$$\sum_{j=1}^{n} a_{kj}x_j = \sum_{j=1}^{n} a_{hj}x_j = 0.$$

Therefore

$$|a_{kk}|\,|x_k| = \left| - \sum_{\substack{j=1 \\ j \neq k}}^{n} a_{kj}x_j \right| \leq \sum_{\substack{j=1 \\ j \neq k}}^{n} |a_{kj}|\,|x_j| \leq \sum_{\substack{j=1 \\ j \neq k}}^{n} |a_{kj}|\,|x_h| = |x_h|P_k,$$

$$|a_{hh}|\,|x_h| = \left| - \sum_{\substack{j=1 \\ j \neq h}}^{n} a_{hj}x_j \right| \leq \sum_{\substack{j=1 \\ j \neq h}}^{n} |a_{hj}|\,|x_j| \leq \sum_{\substack{j=1 \\ j \neq h}}^{n} |a_{hj}|\,|x_k| = |x_k|P_h$$

and

$$|a_{kk}|\,|a_{hh}| \leq P_k P_h$$

contradicting (6).

The second part of (a) is obtained by applying the first part to the matrix A^T.

(b)   Let λ_t be a characteristic root of A. Then

$$\det(\lambda_t I_n - A) = 0$$

and, by part (a), neither

$$|\lambda_t - a_{ii}|\,|\lambda_t - a_{jj}| > P_i P_j$$

nor

$$|\lambda_t - a_{ii}|\,|\lambda_t - a_{jj}| > Q_i Q_j$$

can hold for all i,j ($i \neq j$).

Our next result is the justly famous *Hadamard determinant theorem*. There are many proofs and substantial generalizations of this result extant. We shall state and prove the theorem for nonnegative hermitian matrices. The more general formulation will be stated in Exercise 5.

Theorem 5.7 *If A is an n-square positive semidefinite hermitian matrix then*

$$\det (A) \le \prod_{i=1}^{n} a_{ii}. \tag{10}$$

Equality holds in (10) if and only if A is the diagonal matrix diag $(a_{11}, \ldots, a_{nn})$ *or some a_{ii} is zero.*

Proof. Suppose first that $a_{ii} = 0$ for some i. Then for any $j \ne i$, by Theorem 4.11, $\det (A[i,j|i,j]) \ge 0$, and hence (since $a_{ii} = 0$), $-|a_{ij}|^2 \ge 0$. Thus $a_{ij} = 0$, $j \ne i$. It follows that $A_{(i)} = 0$ and hence both sides of (10) are 0. Thus assume that every $a_{ii} > 0$, $i = 1, \ldots, n$. Then let $B = DAD$, where

$$D = \text{diag } (a_{11}^{-\frac{1}{2}}, \ldots, a_{nn}^{-\frac{1}{2}}).$$

We know that B is nonnegative hermitian and moreover $b_{ii} = 1$, $i = 1, \ldots, n$. Let $\lambda_1, \ldots, \lambda_n$ be the characteristic roots of B. Then, by Theorem 1.5 (a) and the arithmetic-geometric mean inequality,

$$\det (B) = \prod_{i=1}^{n} \lambda_i \le \left[\frac{1}{n} \sum_{i=1}^{n} \lambda_i \right]^n = \left(\frac{\text{tr } (B)}{n} \right)^n \tag{11}$$

$$= \left(\frac{n}{n} \right)^n = 1.$$

Thus

$$1 \ge \det (B) = \det (DAD) = [\det (D)]^2 \det (A)$$

$$= \prod_{i=1}^{n} a_{ii}^{-1} \det (A).$$

Hence

$$\det (A) \le \prod_{i=1}^{n} a_{ii}.$$

The only inequality used to establish (10) occurs in (11) in the application of the arithmetic-geometric mean inequality,

$$\prod_{i=1}^{n} \lambda_i \leq \left[\frac{1}{n} \sum_{i=1}^{n} \lambda_i \right]^n.$$

This can be equality if and only if $\lambda_1 = \cdots = \lambda_n$. Call this common value λ. Then since $n\lambda = \operatorname{tr}(B) = n$, it follows that $\lambda = 1$. Hence B, which is hermitian, must be the identity matrix I_n and thus $A = D^{-2} = \operatorname{diag}(a_{11}, \ldots, a_{nn})$. The proof is over.

Our next result is a generalization of Theorem 5.2. It is known as the *Courant-Fischer theorem* and has all kinds of important applications. We shall see that it can be used to derive interesting inequalities between the characteristic roots of a hermitian matrix and the characteristic roots of a submatrix. This fact can in turn be used to reduce substantially the labor implicit in Theorem 4.11 for verifying positive-definiteness. We state this result for linear transformations.

Theorem 5.8   *Let U be a unitary n-dimensional space and assume that $T \in L(U,U)$ is a hermitian transformation with characteristic roots*

$$\lambda_1 \geq \lambda_2 \geq \cdots \geq \lambda_n.$$

Let $1 \leq k \leq n$ and for each subspace W, $\dim W = n - k + 1$, set

$$c_k(W) = \max_{x \in W, \|x\| = 1} (Tx,x)$$

and

$$d_k(W) = \min_{x \in W, \|x\| = 1} (Tx,x).$$

Then

(a) $c_k(W) \geq \lambda_k, \quad d_k(W) \leq \lambda_{n-k+1}, \qquad k = 1, \ldots, n.$

(b)   *Moreover, for each $k = 1, \ldots, n$, there exist subspaces W_1 and W_2 of U of dimension $n - k + 1$ such that*

$$c_k(W_1) = \lambda_k, \quad d_k(W_2) = \lambda_{n-k+1}.$$

(The notation $\max\limits_{x \in W, \|x\| = 1} (Tx,x)$ *means the maximum value of* (Tx,x) *as* x *runs over unit vectors in* W.)

Proof. Let $u_1, \ldots, u_n$ be an o.n. set of characteristic vectors of T corresponding respectively to $\lambda_1, \ldots, \lambda_n$. If dim $W = n - k + 1$, then, from Theorem 3.3, Chap. 1,

$$
\begin{aligned}
\dim (W \cap \langle u_1, \ldots, u_k \rangle) &= \dim W + \dim \langle u_1, \ldots, u_k \rangle \\
&\quad - \dim (W + \langle u_1, \ldots, u_k \rangle) \\
&\geq n - k + 1 + k - n \\
&= 1.
\end{aligned}
$$

Hence there exists a nonzero vector $u \in W \cap \langle u_1, \ldots, u_k \rangle$ which we may assume is of unit length. Since $u_1, \ldots, u_k$ are o.n. we have, by Theorem 3.4, Chap. 1,

$$
u = \sum_{i=1}^{k} (u,u_i)u_i, \quad \sum_{i=1}^{k} |(u,u_i)|^2 = 1.
$$

Thus, by Theorem 5.1,

$$
\begin{aligned}
c_k(W) &= \max_{x \in W, \|x\| = 1} (Tx,x) \\
&\geq (Tu,u) \\
&= \left(T \sum_{i=1}^{k} (u,u_i)u_i, \sum_{i=1}^{k} (u,u_i)u_i \right) \\
&= \left(\sum_{i=1}^{k} (u,u_i)\lambda_i u_i, \sum_{i=1}^{k} (u,u_i)u_i \right) \\
&= \sum_{i=1}^{k} |(u,u_i)|^2 \lambda_i \\
&\geq \lambda_k.
\end{aligned}
$$

Therefore

$$
c_k(W) \geq \lambda_k.
$$

Similarly, there exists a unit vector $v \in \langle u_n, u_{n-1}, \ldots, u_{n-k+1} \rangle \cap W$, and thus

$$
\begin{aligned}
d_k(W) &= \min_{x \in W, \|x\| = 1} (Tx,x) \\
&\leq (Tv,v) \\
&= \left(T \sum_{i=n-k+1}^{n} (v,u_i)u_i, \sum_{i=n-k+1}^{n} (v,u_i)u_i \right) \\
&= \sum_{i=n-k+1}^{n} |(v,u_i)|^2 \lambda_i \\
&\leq \lambda_{n-k+1}.
\end{aligned}
$$

If we choose $W_1 = \langle u_k, \ldots, u_n \rangle$ and $W_2 = \langle u_1, \ldots, u_{n-k+1} \rangle$ then we can check, by calculations very similar to the ones we just made, that

$$
c_k(W_1) = \lambda_k, \quad d_k(W_2) = \lambda_{n-k+1}.
$$

We can apply Theorem 5.8 to derive some rather extraordinary results about the characteristic roots of a submatrix of a hermitian matrix. These are known as *Cauchy's inequalities*.

Theorem 5.9 *Let A be an n-square hermitian matrix with characteristic roots*

$$
\lambda_1 \geq \cdots \geq \lambda_n.
$$

Let B be a k-square principal submatrix of A with characteristic roots

$$
\mu_1 \geq \cdots \geq \mu_k.
$$

Then

$$
\lambda_s \geq \mu_s \geq \lambda_{n-k+s}, \qquad s = 1, \ldots, k. \tag{12}
$$

Proof.   Consider the two vector spaces $V_n(R)$ and $V_k(R)$, where R is the field of complex numbers, with standard inner products. Now the transformations T_A and T_B defined by $T_A x = Ax$ and $T_B y = By$ are hermitian on their respective spaces. Moreover, the characteristic roots of T_A are $\lambda_1, \ldots, \lambda_n$ and those of T_B are $\mu_1, \ldots, \mu_k$ (see Theorem

4.4). Suppose $B = A[\omega|\omega]$, $\omega \in Q_{k,n}$. Let $y = (y_1, \ldots, y_k) \in V_k(R)$ and let

$$x_y = \sum_{t=1}^{k} y_t e_{\omega_t}.$$

Then clearly

$$(T_B y, y) = (T_A x_y, x_y).$$

Let $y^1, \ldots, y^k$ be an o.n. set of characteristic vectors for T_B corresponding to $\mu_1, \ldots, \mu_k$ respectively. For $1 \leq s \leq k$, let $W_1 = \langle y^k, \ldots, y^s \rangle \subset V_k(R)$. Then, as we saw in Theorem 5.8,

$$\mu_s = \max_{y \in W_1, \|y\| = 1} (T_B y, y).$$

Clearly, as y runs over W_1, x_y runs over a subspace W_1' of dimension $k - s + 1$ in $V_n(R)$. Thus, since $\|y\| = \|x_y\|$, we have

$$\mu_s = \max_{y \in W_1, \|y\| = 1} (T_B y, y)$$

$$= \max_{x_y \in W_1', \|x_y\| = 1} (T_A x_y, x_y).$$

This last expression is the maximum of $(T_A x_y, x_y)$, where x_y runs over the unit vectors in a subspace of $V_n(R)$ of dimension $k - s + 1 = n - (n - k + s) + 1$. Thus, according to Theorem 5.8,

$$\max_{x_y \in W_1', \|x_y\| = 1} (T_A x_y, x_y) \geq \lambda_{n-k+s},$$

or

$$\mu_s \geq \lambda_{n-k+s}.$$

Similarly, let $W_2 = \langle y^1, \ldots, y^s \rangle$ and let W_2' be the totality of vectors x_y for $y \in W_2$. Then clearly dim $W_2' = s$ and

$$\mu_s = \min_{y \in W_2, \|y\| = 1} (T_B y, y)$$

$$= \min_{x_y \in W_2', \|x_y\| = 1} (T_A x_y, x_y).$$

This last expression is the minimum of $(T_A x_y, x_y)$ where x_y runs over the unit vectors in a subspace of $V_n(R)$ of dimension $s = n - (n - s + 1) + 1$. Thus, according to Theorem 5.8, we have as before

$$\min_{x_y \in W'_2, \|x_y\| = 1} (T_A x_y, x_y) \leq \lambda_{n - (n - s + 1) + 1} = \lambda_s.$$

In other words,

$$\mu_s \leq \lambda_s.$$

This completes the proof.

In case the matrix B in Theorem 5.9 is an $(n - 1)$-square principal submatrix of A then the inequalities (12) become

$$\lambda_s \geq \mu_s \geq \lambda_{s+1}, \qquad s = 1, \ldots, n - 1. \tag{13}$$

The inequalities (13) now permit us to improve Theorem 4.11. We have:

Theorem 5.10 *If A is an n-square hermitian matrix then A is positive-definite hermitian if and only if*

$$\det (A[1, \ldots, k|1, \ldots, k]) > 0, \qquad k = 1, \ldots, n. \tag{14}$$

Proof. The implication in one direction has been proved in Theorem 4.11. Now suppose the inequalities (14) hold. We prove the theorem by induction on n. Now, by the induction assumption, $A[1, \ldots, n - 1|1, \ldots, n - 1]$ must be positive-definite because

$$\det (A[1, \ldots, k|1, \ldots, k]) > 0, \qquad k = 1, \ldots, n - 1.$$

Let μ_{n-1} be the smallest characteristic root of $A[1, \ldots, n-1|1, \ldots, n-1]$. We know that $\mu_{n-1} > 0$. According to (13), we can conclude that

$$\lambda_s \geq \mu_{n-1} > 0, \qquad s = 1, \ldots, n - 1. \tag{15}$$

But

$$\det (A) = \Big(\prod_{s=1}^{n-1} \lambda_s \Big) \lambda_n.$$

However, by (15),

$$\prod_{s=1}^{n-1} \lambda_s > 0,$$

and by assumption, det $(A) > 0$. Hence $\lambda_n > 0$. In other words, all characteristic roots of A are positive, and hence A is positive-definite.

Quiz

Answer true or false:

1. If A is an n-square hermitian matrix, then

$$\det (A) \leq \left[\frac{\operatorname{tr} (A)}{n} \right]^n.$$

2. The matrix

$$A = \begin{bmatrix} 10 & 2 & -1 \\ 2 & 5 & 3 \\ -1 & 3 & 7 \end{bmatrix}$$

is positive-definite.

3. The matrix

$$\begin{bmatrix} 3 & 1 & 1 & 2 \\ 1 & -4 & 1 & -1 \\ 0 & -1 & 3 & 1 \\ 1 & 1 & 0 & 5 \end{bmatrix}$$

is nonsingular.

4. If $A = (a_{ij}) \in M_n(R)$ is nonsingular, then

$$|a_{ii}| > \sum_{\substack{t=1 \\ t \neq i}}^{n} |a_{it}|, \qquad i = 1, \ldots, n.$$

5. In each of the Geršgorin disks

$$|z - a_{ii}| \leq P_i, \qquad i = 1, \ldots, n,$$

there is at least one characteristic root of $A = (a_{ij}) \in M_n(R)$.

6. Let λ_1 be a characteristic root of $\begin{bmatrix} a_{11} & a_{12} \\ a_{21} & a_{22} \end{bmatrix}$; then

$$|\lambda_1 - a_{11}|\,|\lambda_1 - a_{22}| \leq |a_{12}|\,|a_{21}|.$$

7. If $H = (h_{ij})$ is a nonsingular hermitian matrix and $h_{11} = 0$ then H has a positive characteristic root.

8. All the characteristic roots of the matrix

$$\begin{bmatrix} -3 & 1 & 1 \\ 2 & -4 & -1 \\ 2 & 2 & -5 \end{bmatrix}$$

have negative real parts.

9. If $A = (a_{ij}) \in M_n(R)$ and

$$a_{ii} = \pm n, \quad i = 1, \ldots, n,$$
$$a_{ij} = \pm 1, \quad i,j = 1, \ldots, n, \quad i \neq j,$$

then $\det(A) \neq 0$ for any choice of $\pm$ signs.

10. If $\lambda_1 \geq \cdots \geq \lambda_n > 0$ and $\mu_1 \geq \cdots \geq \mu_n > 0$ are the characteristic roots of hermitian matrices H and K respectively and if $\lambda_1 > \mu_1$ then for any n-tuple x with positive entries $(Hx,x) > (Kx,x)$.

Exercises

1. Using Theorem 5.2, show that if $\lambda_1 \geq \cdots \geq \lambda_n$ are the characteristic roots of a hermitian matrix $H = (h_{ij})$, then

$$\lambda_n \leq h_{ii} \leq \lambda_1, \quad i = 1, \ldots, n.$$

2. Show that

$$|\lambda_t| \geq \min_i (|a_{ii}| - P_i)$$

for any characteristic root λ_t of $A = (a_{ij})$.

3. Show that

$$|\lambda_t| \leq \max_i \sum_{t=1}^n |a_{it}|$$

for any characteristic root λ_t of $A = (a_{ij})$. Deduce an upper bound for $|\det(A)|$.

4. Find an upper and a lower bound for the real parts of the characteristic roots of the matrix

$$A = \begin{bmatrix} 3 & 2 & 1 \\ 4 & 3 & 6 \\ 5 & 0 & 3 \end{bmatrix}.$$

5. Let A be an n-square complex matrix. Show that

$$|\det (A)| \le \Big(\prod_{i=1}^{n} \sum_{j=1}^{n} |a_{ij}|^2 \Big)^{\frac{1}{2}}.$$

6. Let A be a 3-square hermitian matrix. Assume that $a_{11} > 0$, $\det (A[1,2|1,2]) < 0$, $\det (A) > 0$. Show that A has one positive and two negative characteristic roots.

7. Using Theorem 2.6, or otherwise, prove that

$$\sum_{t=1}^{n} |\operatorname{Re} (\lambda_t)|^2 \le \tfrac{1}{4} n^2 \max_{i,j} |a_{ij} + \bar{a}_{ji}|^2$$

where the λ_t are the characteristic roots of $A = (a_{ij}) \in M_n(R)$.

8. Show that a symmetric nonzero matrix all of whose entries are non-negative has at least one positive characteristic root.

9. Show that the matrix

$$\begin{bmatrix} 1 - 3i & i & 1 + i \\ 1 + i & 2 + 4i & 1 - i \\ 1 - i & -i & 1 + 3i \end{bmatrix}$$

has no real roots.

10. Let A and B be n-square hermitian matrices with characteristic roots $\lambda_1 \ge \cdots \ge \lambda_n$ and $\mu_1 \ge \cdots \ge \mu_n$ respectively. Let $\sigma_1 \ge \cdots \ge \sigma_n$ be the characteristic roots of $A + B$. Then prove that $\max \{\lambda_k + \mu_n, \mu_k + \lambda_n\} \le \sigma_k \le \min \{\lambda_k + \mu_1, \mu_k + \lambda_1\}$.

11. Show that if A is a positive-definite n-square hermitian matrix with characteristic roots $\lambda_1 \ge \cdots \ge \lambda_n > 0$ and B is a k-square principal submatrix of A, then

$$\prod_{j=1}^{k} \lambda_{n-j+1} \le \det (B) \le \prod_{j=1}^{k} \lambda_j.$$

12. Let T be a positive-definite hermitian transformation on a unitary space
     U, dim $U = n$. Let $x_1, \ldots, x_k$ be any k orthonormal vectors in U.
     Let $\lambda_1 \geq \cdots \geq \lambda_n$ be the characteristic roots of T. Show that

$$\prod_{j=1}^{k} \lambda_{n-j+1} \leq \det\left[(Tx_i, x_j)\right] \leq \prod_{j=1}^{k} \lambda_j.$$

Answers and Solutions

CHAPTER 1

Section 1.1

Quiz

1. **True.** We first show that $0x = x0 = 0$ for any x. For, $x(x + 0) = xx + x0$ and $x(x + 0) = xx$, by Def. 1.0 (vi). Thus, $xx = xx + x0$. Now add $-(xx)$ to both sides to obtain $-(xx) + xx = -(xx) + (xx + x0) = x0$ and, finally, $x0 = 0$ from Def. 1.0 (ii), (v). Now, suppose $ab = 0$ and $a \neq 0$. Then, by Def. 1.0 (ix) there exists an element $a^{-1} \in R$ such that $a^{-1}a = 1$. Then, $a^{-1}(ab) = (a^{-1}a)b = 1b = b$, by Def. 1.0 (ii), (viii). Hence, $b = a^{-1}(ab) = a^{-1}0 = 0$. If $b \neq 0$, then we prove $a = 0$ in a similar way.

2. **True.** Suppose $v_k = 0_V$. Let $c_i = 0$, $i \neq k$, $c_k = 1$. Then $c_1 v_1 + \cdots + c_r v_r = c_k v_k = 1 \cdot 0_V$. We prove by an argument that is almost the same as the preceding one that $a0_V = 0_V$ for any scalar a. For, $a0_V = a(0_V + 0_V) = a0_V + a0_V$. Now add $-(a0_V)$ to both sides to obtain $a0_V = 0_V$. We have used Def. 1.1 (v), (vii), (ii). Hence we know that $1 \cdot 0_V = 0_V$ and $v_1, \ldots, v_r$ are linearly dependent.

3. **True.**   Let $v_1, \ldots, v_n$ be a basis and suppose $v_{i_1}, \ldots, v_{i_r}$ is a subset. If $c_{i_1}, \ldots, c_{i_r}$ are scalars, not all zero, such that $c_{i_1}v_{i_1} + \cdots + c_{i_r}v_{i_r} = 0$ then define c_t, $t \neq i_1, \ldots, i_r$, to be 0. Then $c_1v_1 + \cdots + c_nv_n = 0$ and not all of $c_1, \ldots, c_n$ are zero. But $v_1, \ldots, v_n$ is a basis and hence linearly independent.

4. **False.**   Set $u_1 = (1,0,0,0)$, $u_2 = (0,1,0,0)$, $u_3 = (-1,-1,0,0)$ and observe that $u_1 + u_2 + u_3 = 0$. Hence u_1, u_2, u_3 are linearly dependent. But clearly u_i is not a multiple of u_j for $j \neq i$.

5. **False.**   If this set were to be a subspace of $V_2(R)$, then the sum of any two vectors in the set would also have to be in the set. But $(1,1)$ and $(1,-1)$ both satisfy $x_1{}^2 - x_2{}^2 = 0$. However, $(1,1) + (1,-1) = (2,0)$ does not.

6. **False.**   Take V to be $V_2(R)$, R the real number field. Let $u_1 = (1,0)$, $u_2 = (0,1)$ and let $X = \{u_1, u_2\}$. Then $u_1 + u_2 \in \langle u_1, u_2 \rangle$. However, $u_1 + u_2 = (1,1)$ is not in X.

7. **True.**   Let $w_1, \ldots, w_k$ be a maximal set in W. It is linearly independent by definition. Let $x \in W$ and consider the set of $k + 1$ vectors $x, w_1, \ldots, w_k$. These must be linearly dependent because of the maximality of the set $w_1, \ldots, w_k$. Thus there exist scalars $c, d_1, \ldots, d_k$, not all 0, such that $cx + d_1w_1 + \cdots + d_kw_k = 0$. If $c = 0$, then $d_1w_1 + \cdots + d_kw_k = 0$ contradicts the linear independence of $w_1, \ldots, w_k$. Hence $c \neq 0$ and $x = (-c^{-1}d_1)w_1 + \cdots + (-c^{-1}d_k)w_k$. This shows that $w_1, \ldots, w_k$ is a spanning set and hence a basis.

8. **True.**   In general, any subspace W must contain the zero vector because $0w = 0_V$ for any vector $w \in W$ and a scalar multiple of anything in W is in W.

9. **False.**   The zero vector $(0,0)$ must be in any subspace. But $x_1 = 0$, $x_2 = 0$ does not satisfy the equation $3x_1 + 2x_2 = 1$.

10. **False.**   Let $V = V_2(R)$, R the real-number field. Set $X = \{(0,1),(1,0)\}$, $Y = \{(0,2),(2,0)\}$. Then $\langle X \rangle = \langle Y \rangle = V$. But X and Y do not overlap.

Exercises

1. The zero element is a. The multiplicative identity is b. The sum and product of two elements of R is well defined. The associative and distributive laws must be verified by taking all possible cases. (We remark that R is in fact just the field of integers modulo 3.)

2. The zero element is $(0, 0, 0, \ldots)$. The statements (i)–(x) of Def. 1.1 are easy to verify directly. For example, $-(a_0, a_1, \ldots) = (-a_0, -a_1, \ldots)$. The vectors $(1, 0, 0, \ldots)$, $(0, 1, 0, \ldots)$, $(0, 0, 1, 0, \ldots)$, $\ldots$, form a basis which contains an infinite number of vectors.

3. The vectors $u_1 = (1,0,0)$, $u_2 = (0,1,0)$ and $u_3 = (1,1,1)$ are all in X. Also $u_3 - u_1 - u_2 = (0,0,1) \in \langle X \rangle$. Hence $\langle X \rangle$ contains a basis $(1,0,0)$, $(0,1,0)$, $(0,0,1)$ for $V_3(R)$. Thus $\langle X \rangle = V_3(R)$. The vectors u_1, u_2, u_3 constitute a basis for $V_3(R)$ and are in X.

4. If W is a subspace of $V_2(R)$ then, by Theorem 1.3, the possibilities for dim W are 0,1,2. If dim $W = 0$ then, of course, W consists of just the zero vector. If dim $W = 2$, then $W = V_2(R)$ by Theorem 1.3. If dim $W = 1$, then W consists of all multiplies of a fixed vector $w \in W$, $w \neq 0$.

5. The verification that the real numbers are a vector space over the rational numbers is easy. Suppose then that the real numbers were a finite dimensional vector space over the rationals. We shall make this argument depend on the fact that the number π is not a root of a polynomial equation with rational coefficients. This is not an easy fact in itself but it is a widely known one. Suppose then that the dimension of the reals over the rationals is n. Consider the $n + 1$ numbers $\pi, \pi^2, \ldots, \pi^{n+1}$. These must be linearly dependent and hence there exist rational numbers $c_1, \ldots, c_{n+1}$, not all zero, such that $c_1\pi + c_2\pi^2 + \cdots + c_{n+1}\pi^{n+1} = 0$. This is a contradiction.

6. Clearly $0u = 0_V \in W$. Also, if $u \in W$, then $-1u = -u \in W$. If u and v are in W then $1 \cdot u + 1 \cdot v = u + v \in W$. The rest of the conditions (i)–(x) of Def. 1.1 are equally easily verified.

Section 1.2

Quiz

1. **True.** $T^2(au_1 + bu_2) = T[T(au_1 + bu_2)] = T(aTu_1 + bTu_2) = aT(Tu_1) + bT(Tu_2) = aT^2u_1 + bT^2u_2$. Thus $T^2 \in L(U,U)$.

2. **True.** To say that $\eta(T) = \dim U$ is to say that $\ker T = U$. Thus if $u \in U$, then $Tu = 0$. But this means that $T = 0$.

3. **True.** This is a specialization of Theorem 2.1 (ii).

4. **True.** For if $u \in \ker S$ then $Su = 0$. Thus $TSu = T(Su) = T0 = 0$. Hence $u \in \ker TS$.

5. **True.** $T(e_1 + e_2 + e_3) = Te_1 + Te_2 + Te_3 = e_1 + e_3 - e_3 = e_1$. Hence Te_1 cannot be $2e_1$.

6. **True.** This is established in the proof of Theorem 2.1 (ii).

7. **False.** Let $U = V_2(R)$. Define T and S by $Te_1 = e_2$, $Te_2 = e_1$, $Se_1 = e_1$, $Se_2 = e_1 + e_2$. Then $(S + T)^2e_1 = (S + T)(S + T)e_1 = (S + T)(e_1 + e_2) = S(e_1 + e_2) + T(e_1 + e_2) = e_1 + e_1 + e_2 + e_2 + e_1 = 3e_1 + 2e_2$. On the other hand, $(S^2 + 2ST + T^2)e_1 = e_1 + 2(e_1 + e_2) + e_1 = 4e_1 + 2e_2$. In other words, $(S + T)^2 = S^2 + ST + TS + T^2$, but since S and T do not in general commute, it is not the case that $ST = TS$. Thus we cannot conclude that $(S + T)^2 = S^2 + 2ST + T^2$.

8. **True.** By Theorem 2.1 (i), $\rho(T) + \eta(T) = n$. It is generally true that the product of two nonnegative numbers is at most the square of their arithmetic mean. Thus

$$\rho(T)\eta(T) \leq \left(\frac{\rho(T) + \eta(T)}{2}\right)^2 = \left(\frac{n}{2}\right)^2 = \frac{n^2}{4}.$$

9. **False.** Let $U = V_2(R)$ and define $Te_1 = 0$, $Te_2 = e_1$. Then $T^2e_1 = T0 = 0$, $T^2e_2 = TTe_2 = Te_1 = 0$. Thus $T^2 = 0$, but $T \neq 0$.

10. **True.** Since $\ker T \neq 0$ there exists a nonzero vector u_1 such that $Tu_1 = 0$. Also $0 \in \ker T$. Thus we can take $u_2 = 0$.

Exercises

1. We verify for example that $T + (-T) = 0$. Thus $[T + (-T)]u = Tu + (-T)u = Tu - Tu = 0$. The rest of the axioms in Def. 1.1 are equally easily verified.

2. The process of linear extension consists of first assigning values to Tu_s, $s = 1, \ldots, n$, where $u_1, \ldots, u_n$ is a basis of U. Then the definition of the value of T is extended to any vector by

$$Tu = \sum_{s=1}^{n} a_s Tu_s.$$

To see that T is linear we must check that

$$T(av + bw) = aTv + bTw$$

for any scalars a and b and any vectors v and w. Now if

$$v = \sum_{s=1}^{n} c_s u_s \quad \text{and} \quad w = \sum_{s=1}^{n} d_s u_s$$

then

$$T(av + bw) = T\left(a \sum_{s=1}^{n} c_s u_s + b \sum_{s=1}^{n} d_s u_s\right)$$

$$= T\left(\sum_{s=1}^{n} (ac_s + bd_s)u_s\right)$$

$$= \sum_{s=1}^{n} (ac_s + bd_s)Tu_s$$

$$= a \sum_{s=1}^{n} c_s Tu_s + b \sum_{s=1}^{n} d_s Tu_s$$

$$= aTv + bTw.$$

3. Define the linear transformation $A: V_4(R) \rightarrow V_3(R)$ exactly as in the discussion preceding Theorem 2.2. Then $Ae_1 = v_1 = (2,0,0)$, $Ae_2 = v_2 = (1,1,0)$, $Ae_3 = v_3 = (-3,1,1)$, $Ae_4 = v_4 = (0,-7,1)$. Now v_1, v_2, v_3 con-

stitute a basis for $V_3(R)$, and hence $(0,-1,0)$ is in the range of A. In fact, $\frac{1}{2}v_1 - v_2 = (0,-1,0)$.

4. We know from Theorem 2.1 (i) that

$$\rho(T) + \eta(T) = \dim U.$$

Thus

$$\eta(T) = \dim U - \rho(T) > \dim V - \rho(T).$$

Now rng $T \subset V$ and $\rho(T) = \dim$ rng T. Hence $\rho(T) \leq \dim V$. It follows that $\eta(T) \geq 1$, and thus there exists a nonzero $u \in U$ for which $Tu = 0$.

5. Let U and V be vector spaces over a field R. Let 0 be the zero linear transformation in $L(U,V)$ and 0_R the zero in R. Then

$$0(0_R 0_U) = 0_V.$$

6. Clearly, if $S = T$ then $Su_s = Tu_s$. If $Su_s = Tu_s$, $s = 1, \ldots, n$, then

$$S\left(\sum_{s=1}^{n} c_s u_s \right) = \sum_{s=1}^{n} c_s S u_s = \sum_{s=1}^{n} c_s T u_s = T\left(\sum_{s=1}^{n} c_s u_s \right).$$

Any vector is of the form $\sum_{s=1}^{n} c_s u_s$ and hence S and T agree on all of U.

7. Clearly $T \neq 0$ by definition. Also $T^2 e_1 = TT e_1 = Te_2 = e_3$ and hence $T^2 \neq 0$. However, $T^3 e_1 = T^2 e_2 = Te_3 = 0$, $T^3 e_2 = T^2 T e_2 = T^2 e_3 = TT e_3 = T0 = 0$ and $T^3 e_3 = 0$. Thus $T^3 = 0$.

8. The range of T is spanned by $Te_1 = e_2$ and $Te_2 = e_3$ and these vectors are linearly independent. Thus $\rho(T) = 2$ and $\eta(T) = \dim V_3(R) - \rho(T) = 3 - 2 = 1$.

9. This is the system of equations (with the right hand sides all replaced by zeros) considered in (20). We established there that $v_1 = Ae_1 = (2,1,-1)$, $v_2 = Ae_2 = (6,3,-3)$, $v_3 = Ae_3 = (-1,1,-1)$, $v_4 = Ae_4 = (1,0,0)$. Also $v_2 = 3v_1$, $v_3 + 3v_4 = v_1$. Thus $v_2 - 3v_3 - 9v_4 = 0$, or $A(e_2 - 3e_3 - 9e_4) = 0$. But $e_2 - 3e_3 - 9e_4 = (0,1,-3,-9)$. Thus $x_1 = 0$, $x_2 = 1$, $x_3 = -3$, $x_4 = -9$ is a solution.

Section 1.3

Quiz

1. **True.** Let u be any vector. Then $(u,0_V) = (u,0 \cdot 0_V) = 0(u,0_V) = 0$.
2. **True.** Complete u and v to a basis of V and let $(\ ,\ )$ be the standard inner product relative to this basis. Then $(u,u) = 1$, $(v,v) = 1$ and $(u,v) = 0$ by Def. 3.2.

3. **False.** Let $V = V_2(R)$, R the complex number field. Let $y = (1,0)$, $z = (-1,0)$, $x = (0,1)$. Then $(x, ay + bz) = \bar{a}(x,y) + \bar{b}(x,z) = 0$ so $x \in \langle y,z \rangle^{\perp}$. However, $0x + y + z = 0$ so that x,y,z are linearly dependent.

4. **True.** The linearly independent vectors $(2,1,0)$ and $(0,3,2)$ span the plane. We then know from (9) and (10) that the subspace of all vectors perpendicular to the plane is one dimensional.

5. **True.** If $x \in \langle v \rangle$ then $(x,y) = 0$ for all $y \in \langle v \rangle^{\perp}$. Thus $x \in (\langle v \rangle^{\perp})^{\perp}$. On the other hand let $z \in (\langle v \rangle^{\perp})^{\perp}$ and write $z = z_1 + z_2$, where $z_1 \in \langle v \rangle$, $z_2 \in \langle v \rangle^{\perp}$, by using the projection theorem (10) applied to $W = \langle v \rangle$. Then $0 = (z,z_2) = (z_1 + z_2, z_2) = (z_1,z_2) + (z_2,z_2) = \|z_2\|^2$. It follows that $z_2 = 0$ and $z = z_1 \in \langle v \rangle$.

6. **True.** Let $g_1, \ldots, g_n$ be the n linearly independent vectors. They form a basis and hence $v = \sum_{s=1}^{n} c_s g_s$. We are assuming that $(v,g_s) = 0$, $s = 1, \ldots, n$, and hence $\|v\|^2 = (v,v) = \left(v, \sum_{s=1}^{n} c_s g_s \right) = \sum_{s=1}^{n} \bar{c}_s(v,g_s) = 0$. Thus $v = 0$.

7. **False.** Let $V = V_3(R)$, $W_1 = \langle e_1,e_2 \rangle$, $W_2 = \langle e_2,e_3 \rangle$. Then $\dim W_1 \dim W_2 = 4$ while $n^2/4 = 9/4$. Of course if $V = W_1 \dotplus W_2$ then $\dim W_1 + \dim W_2 = n$ and the inequality would indeed be true.

8. **True.** Recall that $\dim (W_1 + W_2) \leq n$. From (9), $\dim (W_1 \cap W_2) = \dim W_1 + \dim W_2 - \dim (W_1 + W_2) > n/2 + n/2 - n = n - n = 0$. Thus the integer $\dim (W_1 \cap W_2)$ is at least 1.

9. **False.** Let $V = V_2(R)$, $W_1 = \langle e_1 \rangle$, $W_2 = \langle e_1 + e_2 \rangle$. If $u \in W_1$ and $v \in W_2$, then $u = ae_1$, $v = b(e_1 + e_2)$ and $(u,v) = (ae_1, b(e_1 + e_2)) = a\bar{b}$. Hence $(u,v) = 0$ only if $a = 0$ or $b = 0$, i.e., only if $u = 0$ or $v = 0$.

10. **False.** Let $n = 2$ and let $x = (1,i)$. Then $(x,x) = 1 \cdot 1 + i \cdot i = 1 + i^2 = 0$. But $x \neq 0$, so the indicated expression fails to satisfy the positive-definite property of the inner product.

Exercises

1. We compute that $x = 2v_1 + (2 - 3i)v_2 - 3iv_3$ and $y = v_1 + (2 + i)v_2 + iv_3$. Then, by the definition of the standard inner product relative to v_1, v_2, v_3, we have $(x,y) = 2 \cdot 1 + (2 - 3i)\overline{(2 + i)} - 3i \cdot \bar{i} = -8i$.

2. (i) $(u,v) = \sum_{i=1}^{n} c_i \bar{d}_i = \overline{\sum_{i=1}^{n} d_i \bar{c}_i} = \overline{(v,u)}$.

   (ii) If $u_1 = \sum_{i=1}^{n} c_i v_i$, $u_2 = \sum_{i=1}^{n} c'_i v_i$, then

$$(au_1 + bu_2, v) = \sum_{i=1}^{n} (ac_i + bc'_i)\bar{d}_i = a \sum_{i=1}^{n} c_i \bar{d}_i + b \sum_{i=1}^{n} c'_i \bar{d}_i$$
$$= a(u_1,v) + b(u_2,v).$$

3. If $\sum\limits_{i=1}^{n} c_i v_i = 0$, then $0 = \left(\sum\limits_{i=1}^{n} c_i v_i, v_j\right) = \sum\limits_{i=1}^{n} c_i(v_i, v_j) = c_j\|v_j\|^2$. But $\|v_j\|^2$
    $\neq 0$ and hence $c_j = 0$, $j = 1, \ldots, n$.

4. Let $u_i = a_i^{\lambda/2}$, $v_i = a_i^{(1-\lambda)/2}$, $i = 1, \ldots, n$. From (24) we have

$$\left|\sum_{i=1}^{n} u_i \bar{v}_i\right|^2 \leq \sum_{i=1}^{n} |u_i|^2 \sum_{i=1}^{n} |v_i|^2$$

or

$$\left(\sum_{i=1}^{n} a_i^{1/2}\right)^2 \leq \sum_{i=1}^{n} a_i^{\lambda} \sum_{i=1}^{n} a_i^{1-\lambda}.$$

Equality holds if and only if there exists a constant c (which must be positive) such that $a_i^{\lambda/2} = c a_i^{(1-\lambda)/2}$, $i = 1, \ldots, n$, i.e., $a_i^{(2\lambda-1)/2} = c$, $i = 1, \ldots, n$. This can happen if and only if $\lambda = \frac{1}{2}$ or $a_1 = \cdots = a_n$.

5. We use the inequality $ab \leq (a^2 + b^2)/2$ for real numbers which follows from $(a - b)^2 \geq 0$. Let $e_1, \ldots, e_n$ be an o.n. basis relative to (,).

Let $E = \max\limits_{i,j} |(e_i, e_j)_1|$. If $u = \sum\limits_{i=1}^{n} c_i e_i$, then

$$(u, u)_1 = \sum_{i,j=1}^{n} c_i \bar{c}_j (e_i, e_j)_1 = \left|\sum_{i,j=1}^{n} c_i \bar{c}_j (e_i, e_j)_1\right|$$

$$\leq \sum_{i,j=1}^{n} |c_i|\,|c_j|\,|(e_i, e_j)_1| \leq E \sum_{i,j=1}^{n} |c_i|\,|c_j|$$

$$\leq E \sum_{i,j=1}^{n} \frac{|c_i|^2 + |c_j|^2}{2} = En \sum_{i=1}^{n} |c_i|^2 = En(u, u).$$

6. In the discussion immediately preceding Theorem 3.4 take W to be $\langle v \rangle^\perp$. Then an o.n. basis for $(\langle v \rangle^\perp)^\perp$ is clearly just $\dfrac{v}{\|v\|}$. Hence the distance from v to $\langle v \rangle^\perp$ is (see p. 39)

$$\left|\left(v, \frac{v}{\|v\|}\right)\right| = \|v\|^2/\|v\| = \|v\|.$$

7. The vectors $u_1 = (1, 2, 2)$ and $u_2 = (3, 1, 2)$ form a basis for the plane. We complete these vectors to a basis of $V_3(R)$ with the vector $u_3 = (1, 1, 0)$. If we apply the Gram-Schmidt process (Theorem 3.2) to u_1, u_2, u_3 we

obtain the o.n. basis $v_1 = \frac{1}{3}(1,2,2)$, $v_2 = \frac{1}{\sqrt{5}}(2,-1,0)$, $v_3 = \frac{1}{3\sqrt{5}}(2,4,-5)$. Then if $u = (1,3,1)$ we compute that $|(u,v_3)| = \frac{3}{\sqrt{5}}$, the required distance.

8. We verify the three properties in Def. 3.1.

(i) $(u,v) = \sum_{i=1}^{n} \lambda_i u_i \bar{v}_i = \overline{\sum_{i=1}^{n} \lambda_i v_i \bar{u}_i} = \overline{(v,u)}.$

(ii) $(au + bw, v) = \sum_{i=1}^{n} \lambda_i (au_i + bw_i)\bar{v}_i = a \sum_{i=1}^{n} \lambda_i u_i \bar{v}_i + b \sum_{i=1}^{n} \lambda_i w_i \bar{v}_i = a(u,v) + b(w,v).$

(iii) $(u,u) = \sum_{i=1}^{n} \lambda_i u_i \bar{u}_i = \sum_{i=1}^{n} \lambda_i |u_i|^2 \geq 0$. Also $(u,u) = 0$ if and only if $u_i = 0$, $i = 1, \ldots n$, (the λ_i are positive).

9. The problem is to verify the properties in Def. 3.1.

(i) $(S,T) = \sum_{i=1}^{n} (Sv_i, Tv_i) = \sum_{i=1}^{n} \overline{(Tv_i, Sv_i)} = \overline{(T,S)}.$

(ii) $(aS_1 + bS_2, T) = \sum_{i=1}^{n} ((aS_1 + bS_2)v_i, Tv_i) = \sum_{i=1}^{n} \{a(S_1 v_i, Tv_i) + b(S_2 v_i, Tv_i)\} = a \sum_{i=1}^{n} (S_1 v_i, Tv_i) + b \sum_{i=1}^{n} (S_2 v_i, Tv_i) = a(S_1, T) + b(S_2, T).$

(iii) $(S,S) = \sum_{i=1}^{n} \|Sv_i\|^2 \geq 0$ with equality if and only if $Sv_i = 0$, $i = 1, \ldots, n$. But $v_1, \ldots, v_n$ form a basis and hence $(S,S) = 0$ if and only if $S = 0$.

10. As in Theorem 2.1 (ii), we define n^2 linear transformations T_{ij} in $L(V,V)$ by $T_{ij}(v_s) = \delta_{is} v_j$, $i,j = 1, \ldots, n$. It was proved there that the T_{ij} are a basis of $L(V,V)$. The problem is to show it is an o.n. basis.

Thus $(T_{pq}, T_{kr}) = \sum_{i=1}^{n} (T_{pq} v_i, T_{kr} v_i) = \sum_{i=1}^{n} (\delta_{pi} v_q, \delta_{ki} v_r) = (v_q, v_r) \sum_{i=1}^{n} \delta_{pi} \delta_{ki}$. Now, unless there is an i such that $i = p$ and $i = k$, this last summation is zero. Thus the last summation is equal to δ_{pk}. Also $(v_q, v_r) = \delta_{qr}$. Hence (T_{pq}, T_{kr}) is 1 or 0 according as $p = k$ and $q = r$ or not.

Section 1.4

Quiz

1. **False.** Let $V = V_2(R)$ and $G = \{e_1, e_2\}$, $H = \{e_1, -e_2\}$. Then $I_V e_1 = e_1$, $I_V e_2 = e_2 = -(-e_2)$. Thus $[I_V]_G^H = \begin{bmatrix} 1 & 0 \\ 0 & -1 \end{bmatrix} \neq I_2$. What is true here is that $[I_V]_G^G$ is always the identity matrix for any vector space V and any basis G.

2. **False.** Since dim $V = 4$ any matrix representation of T must have four rows.

3. **True.** Let G and H be the standard bases in $V_2(R)$ and $V_3(R)$ respectively. Then we compute that

$$A = [T]_G^H = \begin{bmatrix} 1 & -1 \\ 2 & -1 \\ 1 & 0 \end{bmatrix} \quad \text{and hence} \quad A^* = \begin{bmatrix} 1 & 2 & 1 \\ -1 & -1 & 0 \end{bmatrix}.$$

   Thus $T^*(x_1, x_2, x_3) = x_1 T^* e_1 + x_2 T^* e_2 + x_3 T^* e_3 = x_1(1, -1) + x_2(2, -1) + x_3(1, 0) = (x_1 + 2x_2 + x_3, -x_1 - x_2)$.

4. **False.** Take $A = \begin{bmatrix} 0 & 1 \\ 1 & 0 \end{bmatrix}$ and $B = \begin{bmatrix} 1 & 0 \\ 0 & -1 \end{bmatrix}$. Then $AB = \begin{bmatrix} 0 & -1 \\ 1 & 0 \end{bmatrix}$ is not hermitian. If, however, A and B were to commute then $(AB)^* = B^* A^* = BA = AB$ and it would follow that AB is hermitian.

5. **True.** If $q(u) = (Tu, u)$ then $\overline{q(u)} = (u, Tu) = (T^*u, u)$. But $T^* = T$ and hence $\overline{q(u)} = q(u)$ and $q(u)$ is real.

6. **False.** Let $n = 2$, $a_{11} = a_{22} = 0$ and $a_{12} = 1$, $a_{21} = -1$. Then $\varphi(x) = x_1 x_2 - x_2 x_1 = 0$ but $A \neq 0$.

7. **False.** Let $U = V_2(R)$ where R is the complex number field. Take $G = \{e_1, e_2\}$ and use the standard inner product. Define T by $Te_1 = e_2$, $Te_2 = e_1$. Then $q(e_1) = (Te_1, e_1) = (e_2, e_1) = 0$ and similarly $q(e_2) = 0$. However, $T \neq 0$.

8. **False.** Take $A = \begin{bmatrix} 1 & 0 \\ 0 & 0 \end{bmatrix}$ and $B = \begin{bmatrix} 0 & 0 \\ 0 & 1 \end{bmatrix}$.

9. **False.** I_n is unitary but $I_n - I_n = 0$ is not unitary.

10. **True.** $\|Tu\|^2 = (Tu, Tu) = (T^*Tu, u) = 0$. Thus $Tu = 0$ for any u and $T = 0$.

11. **False.** Let $V = V_2(R)$, R the complex numbers. Let $G = \{e_1, e_1 + e_2\}$. Then $[I_V]_G^E = \begin{bmatrix} 1 & 1 \\ 0 & 1 \end{bmatrix} = A$. The matrix A is not normal.

12. **True.** $0 = A^2 = AA^*$. Hence $0 = (AA^*)_{ii} = \sum_{j=1}^{n} a_{ij}\overline{a_{ij}} = \sum_{j=1}^{n} |a_{ij}|^2$, $i = 1, \ldots, n$. It follows that $a_{ij} = 0$, $i, j = 1, \ldots, n$.

13. **True.** As in the preceding solution $A^2 + B^2 = AA^* + BB^*$ and hence

$$\sum_{j=1}^{n} |a_{ij}|^2 + |b_{ij}|^2 = 0, i = 1, \ldots, n.$$ It follows that $A = B = 0$.

14. **False.** Take $V = V_2(R)$, R the real numbers, and use the standard inner product. Define $Te_1 = e_2$, $Te_2 = -e_1$ and observe that $q(u) = (Tu, u) = (T(u_1, u_2), (u_1, u_2)) = ((-u_2, u_2), (u_1, u_2)) = u_2 u_1 - u_1 u_2 = 0$.

15. **True.** This is the same kind of computation as used in the solution of Quiz Problem 12.

Exercises

1. Let i be fixed. Let $r_i = 1$ and $r_j = 0$, $j \neq i$, $j = 1, \ldots, n$. Let $D = \text{diag}(r_1, \ldots, r_n)$. The (i,j) entry of DA is a_{ij}. The (i,j) entry of AD is 0 for $i \neq j$. If $DA = AD$, then $a_{ij} = 0$, $i \neq j$.

2. Let $g_1, \ldots, g_r$ be a basis of W and complete with $g_{r+1}, \ldots, g_n$ to a basis of U. Then $Tg_i \in \langle g_1, \ldots, g_r \rangle = W$, for each $i = 1, \ldots, r$.

   Hence $Tg_j = \sum_{s=1}^{r} a_{sj} g_s$, $j = 1, \ldots, r$. This means that $a_{ij} = 0$, $i = r+1, \ldots, n$, $j = 1, \ldots, r$.

3. We prove $(AB)^* = B^* A^*$ as an example of the technique. $((AB)^*)_{ij} = \overline{(AB)_{ji}} = \overline{\sum_{s=1}^{n} a_{js} b_{si}} = \sum_{s=1}^{n} \bar{a}_{js} \bar{b}_{si} = \sum_{s=1}^{n} (B^*)_{is} (A^*)_{sj} = (B^* A^*)_{ij}$.

4. Choose G to be an o.n. basis such that $g_1, \ldots, g_r$ is a basis of W. Let $A = [T]_G^G$. Then $[T^*]_G^G = A^*$ and since $a_{ij} = 0$, $i = r+1, \ldots, n$, $j = 1, \ldots, r$, it follows that $(A^*)_{ij} = 0$, $j = r+1, \ldots, n$, $i = 1, \ldots, r$. Thus $T^* g_k \in \langle g_{r+1}, \ldots, g_n \rangle$, $k = r+1, \ldots, n$. In other words, $\langle g_{r+1}, \ldots, g_n \rangle$ is an invariant subspace of T^*. But $W = \langle g_1, \ldots, g_r \rangle$ and $W^\perp = \langle g_{r+1}, \ldots, g_n \rangle$.

5. If A and B are unitary matrices, then $(AB)^*(AB) = B^*(A^*A)B = B^*B = I_n$. Similarly, $(AB)(AB)^* = I_n$.

6. The arguments here are just repetitions of the ones already used in the proof of Theorem 4.4.

7. Let $D = A - B$. Then the problem is to show that if $\varphi(x) = \sum_{i,j=1}^{n} d_{ij} x_i \bar{x}_j = 0$ for all $x \in V_n(R)$ then $D = 0$. But this is precisely what is proved in Theorem 4.5.

8. We use the decomposition (30), $T = H + iK$, where H and K are hermitian. Then $q(u) = (Hu, u) + i(Ku, u)$. It follows that if $q(u)$ is real for all u then $(Ku, u) = 0$ for all u. Hence by Theorem 4.5, $K = 0$, and thus $T = H$ is hermitian.

9. $TT^* = H^2 + K^2 + i(KH - HK)$, whereas $T^*T = H^2 + K^2 + i(HK - KH)$. Thus $TT^* = T^*T$ if and only if $KH - HK = HK - KH$, i.e., if and only if $HK = KH$.

10. Let $T = H + iK$, $HK = KH$. Then $T^*T = H^2 + K^2$. Now $0 = (T^2T^*u,Tu) = (TT^*u,T^*Tu) = (T^*Tu,T^*Tu) = \|T^*Tu\|^2$. Thus $0 = T^*T = H^2 + K^2$. It follows from Quiz Problem 13 that $H = K = 0$. Hence $T = 0$.

11. Let $S = T^*$. Then by (23) we have $(u,Sv) = (S^*u,v)$ for all $u \in U$, $v \in V$. But $(u,Sv) = (u,T^*v) = (Tu,v)$ for all $u \in U, v \in V$. Thus $(Tu,v) = (S^*u,v)$ for all $u \in U$, $v \in V$. It follows that $((T - S^*)u, v) = 0$ and setting $v = (T - S^*)u$ we have $T - S^* = 0$.

Section 1.5

Quiz

1. **True.** Suppose $\begin{bmatrix} 1 & 1 \\ 1 & 1 \end{bmatrix} \begin{bmatrix} a & b \\ c & d \end{bmatrix} = \begin{bmatrix} 1 & 0 \\ 0 & 1 \end{bmatrix}$. Then $a + c = b + d = 1$, while $b + d = a + c = 0$, an impossibility.

2. **True.** $\eta(T) = \dim \ker T$, and we apply Theorem 5.1 (b).

3. **False.** If $T = I_U$ and S is singular, then $TS = S$ is singular but I_U is nonsingular.

4. **True.** If $C = \begin{bmatrix} 1 & 1 \\ 3 & 1 \end{bmatrix}$, then $C^{-1} \begin{bmatrix} 1 & 2 \\ 3 & 4 \end{bmatrix} C = \begin{bmatrix} 4 & 2 \\ 3 & 1 \end{bmatrix}$.

5. **True.** If A is unitary, then $AA^* = I_n$; if A is hermitian, $A^* = A$. Hence $A^2 = I_n$.

6. **True.** If A is singular then some $\alpha_i = 0$, otherwise we could set $A^{-1} = \mathrm{diag}\,(\alpha_1^{-1}, \ldots, \alpha_n^{-1})$. Conversely, if some $\alpha_i = 0$ then $(AB)_{(i)} = 0$ for any B. It follows that AB can never be I_n.

7. **False.** Let $U = V_2(R)$, $G = \{e_1,e_2\}$, $H = \{-e_2,e_1\}$. Then $[I_U]_G{}^H = \begin{bmatrix} 0 & -1 \\ 1 & 0 \end{bmatrix}$. But $\begin{bmatrix} 0 & -1 \\ 1 & 0 \end{bmatrix}^{-1} = \begin{bmatrix} 0 & 1 \\ -1 & 0 \end{bmatrix}$.

8. **True.** For if $A_{(i)} = 0$ then $(AB)_{(i)} = 0$ for any B, and hence AB can never be I_n.

9. **True.** $\|Tu\| = 0$ implies $Tu = 0$. Since T is nonsingular, it follows that $T^{-1}(Tu) = u = 0$.

10. **False.** Take $B = \begin{bmatrix} 0 & 1 \\ -1 & 0 \end{bmatrix}$ and $A = \begin{bmatrix} 1 & 2 \\ 2 & -1 \end{bmatrix}$. Then $AB + BA = 0$, but both A and B are nonsingular over the real-number field.

Exercises

1. It follows from $\rho(T) < \dim V$ that rng T is properly included in V (Theorem 1.3). Thus T is not onto V (Theorem 5.1).

2. We have for T^{-1} that $T^{-1}(x_1 + x_2, x_2 + x_3, x_3 + x_1) = (x_1,x_2,x_3)$. Let $y_1 = x_1 + x_2, y_2 = x_2 + x_3, y_3 = x_3 + x_1$, and then

$$x_1 = \frac{y_1 - y_2 + y_3}{2}, \quad x_2 = \frac{y_1 + y_2 - y_3}{2}, \quad x_3 = \frac{-y_1 + y_2 + y_3}{2}.$$

Hence

$$T^{-1}(y_1,y_2,y_3) = \left(\frac{y_1 - y_2 + y_3}{2}, \frac{y_1 + y_2 - y_3}{2}, \frac{-y_1 + y_2 + y_3}{2}\right).$$

3. For, if $\sum_{i=1}^{r} c_i Tg_i = 0$, then

$$0 = T^{-1} \sum_{i=1}^{r} c_i Tg_i = \sum_{i=1}^{r} c_i T^{-1} Tg_i = \sum_{i=1}^{r} c_i g_i.$$

The linear independence of $g_1, \ldots, g_r$ implies that $c_1 = \cdots = c_r = 0$.

4. If ker T were 0 then from Theorem 2.1 (i) we would know that $\rho(T) =$ dim U and hence T would be onto U. But then T would be nonsingular.

5. A is a unit if and only if there exist integers b_1, b_2, b_3, b_4 such that

$$\begin{bmatrix} a_1 & a_2 \\ a_3 & a_4 \end{bmatrix} \begin{bmatrix} b_1 & b_2 \\ b_3 & b_4 \end{bmatrix} = \begin{bmatrix} b_1 & b_2 \\ b_3 & b_4 \end{bmatrix} \begin{bmatrix} a_1 & a_2 \\ a_3 & a_4 \end{bmatrix} = \begin{bmatrix} 1 & 0 \\ 0 & 1 \end{bmatrix}.$$

If we equate entries and set $k = a_1a_4 - a_2a_3$ we find that $k(b_1b_4 - b_2b_3) = 1$. Hence $k = \pm 1$. Conversely, if $k = \pm 1$ we can set $b_1 = ka_4$, $b_2 = -ka_2$, $b_3 = -ka_3$, $b_4 = ka_1$.

6. Suppose that $\sum_{j=1}^{r} d_j h_j = 0$. Then $\sum_{i=1}^{r} \left(\sum_{j=1}^{r} c_{ij}d_j\right) g_i = 0$. Hence $\sum_{j=1}^{r} c_{ij}d_j = 0$, $i = 1, \ldots, r$, or $Cd = 0$, where $C = (c_{ij})$, $d = (d_1, \ldots, d_r) \in V_r(R)$. Now C^{-1} exists, so $0 = C^{-1}Cd = I_r d = d$, and thus $d_1 = \cdots = d_r = 0$.

7. Define $T \in L(U,U)$ by $Tg_j = h_j = \sum_{i=1}^{n} c_{ij}g_i$. If we set $G = \{g_1, \ldots, g_n\}$ and $H = \{h_1, \ldots, h_n\}$, then $[T]_G^H = I_n$ and $[T]_G^G = C$. By Theorem 5.3 applied to $[T]_G^H$ we know that T is nonsingular. Applying the same result to $[T]_G^G = C$, we conclude that C is a unit.

8. Since T is nonsingular, we know that the vectors $h_i = Tg_i$, $i = 1, \ldots, n$, form a basis of U by Theorem 5.1 (d). Set $H = \{h_1, \ldots, h_n\}$ and then $[T]_G^H = I_n$.

9. If $Tu = (1,1,1)$ then $u = T^{-1}(1,1,1)$. If we use the formula for T^{-1} obtained in the solution to Exercise 2 we have $u = (\frac{1}{2},\frac{1}{2},\frac{1}{2})$.

10. No. Because $A_{(n)} = 0$ and hence A must be singular. On the other hand, one can check that $B^T = B^{-1}$ and hence B is nonsingular. Apply Theorem 5.3.

11. Since C^{-1} exists, $0 = C^{-1}Cd = I_n d = d$.

CHAPTER **2**

Section 2.1

Quiz

1. **False.** Let $A = \text{diag}\ (0,1,1)$ and $B = C = \text{diag}\ (1,0,0)$. Then $\rho(A) = 2 > \rho(B) = 1$, but $\rho(AC) = 0 < \rho(BC) = 1$.

2. **False.** Let

$$A = \begin{bmatrix} 0 & 1 & 0 & 0 \\ 0 & 0 & 0 & 0 \\ 0 & 0 & 0 & 1 \\ 0 & 0 & 0 & 0 \end{bmatrix}.$$

Then $\rho(A) = 2$, while $A^2 = 0$ and thus $\rho(A^2) + 1 = 1$.

3. **True.** Let $A,B \in M_{m,n}(R)$. The row spaces $R(A + B)$, $R(A)$, and $R(B)$ consist of all linear combinations of the rows of $A + B$, A and B respectively. Therefore,

$$R(A + B) \subset R(A) + R(B)$$

and, by Theorem 1.3 in Chap. 1,

$$\dim\ [R(A + B)] \leq \dim\ [R(A) + R(B)].$$

Now, by Theorem 3.3 in Chap. 1,

$$\dim\ [R(A) + R(B)] \leq \dim\ [R(A)] + \dim\ [R(B)]$$

and the result follows.

4. **False.** If $A = \text{diag}\ (1,0,0)$ and $B = \text{diag}\ (0,1,1)$, then $\rho(A) < \rho(B)$ but $\rho(A) > \rho(AB)$.

5. **True.** If $\ker T = 0$ then, by Sylvester's law of nullity [Theorem 2.1 (i) in Chap. 1], $\rho(T) = \dim U$. Also, by Theorem 1.6, $\rho(T) = \rho([T]_G{}^H)$.

6. **True.** This follows from Question 5, since $\ker I_U = 0$.

7. **True.** $Tu = 0$ for all $u \in U$ implies that $T = 0$ and therefore, by Theorem 1.6, $\rho([T]_G{}^H) = \rho(T) = 0$.

8. **True.** If $B = 0$ then $AB = 0$ and $\rho(AB) = \rho(A)\rho(B) = 0$. If $\rho(B) \geq 1$ then $\rho(A) \leq \rho(A)\rho(B)$ and since, by Theorem 1.4, $\rho(AB) \leq \rho(A)$, we have $\rho(AB) \leq \rho(A)\rho(B)$.

9. **True.** Clearly $T^n g_j = 0$, $j = 1, \ldots, n$. Therefore $T^n = 0$ and $\rho([T^n]_H{}^H) = 0$. Now, by (12) in Sect. 1.4, $[T^n]_H{}^H = ([T]_H{}^H)^n$.

10. **False.** Let

$$A = B = \begin{bmatrix} 1 & 0 \\ 0 & 0 \end{bmatrix}. \quad \text{Then} \quad AB = \begin{bmatrix} 1 & 0 \\ 0 & 0 \end{bmatrix}$$

and $\rho(AB) = \rho(A) = \rho(B) = 1 > 0$. Nevertheless, both A and B are singular.

Exercises

1. This result follows immediately from Def. 4.1 in Chap. 1. For, $T_{ji}(u_s) = 0$ for $s \neq j$ and $T_{ji}(u_j) = v_i$ and therefore all the entries of $[T_{ji}]_G{}^H$ are 0 except the (i,j) entry which is 1.

2. Let $A = (a_{ij})$ be any matrix in $M_n(R)$. Then

$$A = \sum_{i,j=1}^{n} a_{ij}E_{ij}$$

and therefore $M_n(R) = \langle E_{ij}, \ i,j = 1, \ldots, n \rangle$. Now, if b_{ij} are any scalars then the (i,j) entry in $\sum_{i,j=1}^{n} b_{ij}E_{ij}$ is precisely b_{ij} and therefore $\sum_{i,j=1}^{n} b_{ij}E_{ij} = 0$ implies that $b_{ij} = 0$, $i,j = 1, \ldots, n$. Hence the E_{ij}, $i,j = 1, \ldots, n$, are linearly independent and they form a basis of $M_n(R)$. We now compute the (i,j) entry in the matrix $E_{pq}E_{rs}$:

$$(E_{pq}E_{rs})_{ij} = \sum_{t=1}^{n} (E_{pq})_{it}(E_{rs})_{tj}$$

$$= \sum_{t=1}^{n} \delta_{ip}\delta_{qt}\delta_{rt}\delta_{sj}$$

$$= \delta_{qr}\delta_{ip}\delta_{sj}.$$

Thus

$$E_{pq}E_{rs} = \delta_{qr}E_{ps}.$$

Now, let $A = (a_{ij})$ and $B = (b_{ij})$ be any matrices in $M_n(R)$. Then

$$AB = \left(\sum_{p,q=1}^{n} a_{pq}E_{pq} \right) \left(\sum_{r,s=1}^{n} b_{rs}E_{rs} \right)$$

$$= \sum_{p,q,r,s=1}^{n} a_{pq}b_{rs}E_{pq}E_{rs}$$

$$= \sum_{p,q,r,s=1}^{n} a_{pq}b_{rs}\delta_{qr}E_{ps}$$

$$= \sum_{p,s=1}^{n} \left(\sum_{r=1}^{n} a_{pr}b_{rs} \right) E_{ps}.$$

3. The rank of the matrix $E_{ij} - E_{hk}$ is equal to the number of linearly independent rows in the matrix. The matrix has two distinct rows if $\delta_{ih} = 0$; these rows are linearly dependent or independent according as $\delta_{jk} = 1$ or $\delta_{jk} = 0$. The formula therefore holds in these cases. If $\delta_{ih} = 1$ and $\delta_{jk} = 0$ then $E_{ij} - E_{hk}$ has one nonzero row and $\rho(E_{ij} - E_{hk}) = 1$. Finally, if $\delta_{ih} = \delta_{jk} = 1$ then $E_{ij} - E_{hk} = 0$.

4. In general, $T^{h+1}(x) = T^h(T(x))$, $x \in V$, and therefore rng $T^{h+1} \subset$ rng T^h for all h. Now, since it is given that $\rho(T^{k+1}) = \rho(T^k)$, we have

$$\text{rng } T^{k+1} = \text{rng } T^k.$$

Therefore

$$
\begin{aligned}
\text{rng } T^{k+2} &= \{y \in V | y = T^{k+2}x,\ x \in V\} \\
&= \{y \in V | y = Tz,\ z \in \text{rng } T^{k+1}\} \\
&= \{y \in V | y = Tz,\ z \in \text{rng } T^k\} \\
&= \text{rng } T^{k+1}.
\end{aligned}
$$

We can extend this result, by an obvious inductive argument, and obtain

$$\text{rng } T^{k+t} = \text{rng } T^k,$$

for any positive integer t, and therefore

$$\text{rng } T^i = \text{rng } T^j,$$

for any $i \geq k$ and $j \geq k$.

5. We claim that $\rho(A^k) = \rho(A^{k+1})$ for some $k \leq n$. Indeed, if A is nonsingular then, by Theorem 1.5,

$$\rho(A^2) = \rho(A^{-1}A^2) = \rho(A) = \rho(A^{-1}A) = \rho(I_n) = n.$$

If A is singular, then $\rho(A) < n$ (see Theorem 5.1, Chap. 1, and Theorem 1.6). Therefore, by Theorem 1.4,

$$n > \rho(A) \geq \rho(A^2) \geq \cdots \geq \rho(A^n) \geq 0.$$

If all inequalities are strict then $\rho(A^n) = 0$ and therefore $\rho(A^{n+1}) = 0$. Otherwise, there exists $k < n$ such that $\rho(A^k) = \rho(A^{k+1})$. Now let $T: V_n(R) \to V_n(R)$ be defined by $T(x) = Ax$. Clearly $A = [T]_G^G$, where G is the standard basis in $V_n(R)$. Hence $\rho(A^k) = \rho(T^k)$, $\rho(A^{k+1}) = \rho(T^{k+1})$ and the result follows by Exercise 4.

6. If $\rho(A) < n$ then the dimension of the null space of A is positive. Let B be any matrix whose columns are in the null space of A and are not all zero. Then $AB = 0$ and $B \neq 0$.

7. Let $\{e_1,e_2,e_3,e_4\}$ and $\{e'_1,e'_2,e'_3,e'_4,e'_5,e'_6\}$ denote the standard bases in $V_4(R)$ and $V_6(R)$ respectively. Then

$$\begin{aligned} T(e_1) &= e'_1 + e'_2 + e'_3, \\ T(e_2) &= -e'_1 + e'_4 + e'_5, \\ T(e_3) &= -e'_2 - e'_4 + e'_6, \\ T(e_4) &= -e'_3 - e'_5 - e'_6. \end{aligned}$$

Therefore

$$[T]_G{}^H = \begin{bmatrix} 1 & -1 & 0 & 0 \\ 1 & 0 & -1 & 0 \\ 1 & 0 & 0 & -1 \\ 0 & 1 & -1 & 0 \\ 0 & 1 & 0 & -1 \\ 0 & 0 & 1 & -1 \end{bmatrix}$$

The columns of this matrix add up to 0 and thus they are linearly dependent. Clearly the first three columns are linearly independent (as are the first three rows). Hence $\rho([T]_G{}^H) = 3$ and therefore $\rho(T) = 3$.

8. If $\rho(A) = 1$ then the dimension of the row space of A is 1 and thus every row of A is a scalar multiple of a nonzero row of A; let this row be denoted by y. Then

$$A_{(j)} = x_j y, \qquad x_j \in R, \qquad j = 1, \ldots , n.$$

Now set $x = [x_1, \ldots , x_m]^T$. Then $A = xy$. Reverse the argument to prove the converse.

9. Let $x = [x_1, \ldots , x_m]^T$ and $y = [y_1, \ldots , y_n]$. Set $x_t = \delta_{it}$, $t = 1$, $\ldots , m$, and $y_s = \delta_{js}$, $s = 1, \ldots , n$. Then the (p,q) entry of the matrix xy satisfies

$$(xy)_{pq} = x_p y_q = \delta_{ip}\delta_{jq},$$

and

$$xy = E_{ij}.$$

10. Let $x, y \in V_n(R)$ be in the null space of A, i.e., $Ax = Ay = 0$. Let a, b be any scalars in R. We assert that $ax + by$ is in the null space of A. For,

$$A(ax + by) = aAx + bAy = 0.$$

Hence the null space of A forms a subspace of $V_n(R)$.

Section 2.2

Quiz

1. **False.** If the scalar is 0 the rank of the resulting matrix may be diminished by 1, e.g., if the rows of A are linearly independent then multiplication of any row by 0 diminishes the rank.

2. **True.** Denote the space on the left-hand side by U and the space on the right-hand side by V. Since every vector spanning U is a linear combination of vectors spanning V, we have $U \subset V$. Now, observe that $A_{(2)} = (A_{(1)} + A_{(2)}) - A_{(1)}$ and therefore every vector spanning V is a linear combination of vectors spanning U. Thus $V \subset U$ and it follows that $U = V$.

3. **False.** If the rows of A are linearly independent, then the dimension of the space on the left-hand side is $m - 1$, while that of the space on the right-hand side is m.

4. **True.** Note that $A_{(1)} = \frac{1}{2}(A_{(1)} + A_{(2)}) + \frac{1}{2}(A_{(1)} - A_{(2)})$, $A_{(2)} = \frac{1}{2}(A_{(1)} + A_{(2)}) - \frac{1}{2}(A_{(1)} - A_{(2)})$ and use the same argument as in Question 2.

5. **True.** This follows immediately from Def. 2.6.

6. **True.** See the definition of E_{ij} in the proof of Theorem 2.1.

7. **True.** See the definition of $E_{(i),(j)}$ and formulas (8) and (5).

8. **False.** In fact, by formulas (9) and (6), $(E^{(i) + c(j)})^{-1} = (E_{(j) + c(i)})^{-1} = E_{(j) - c(i)}$.

9. **False.** The matrix A may well be singular and A^{-1} may not exist, e.g., if $B = C = D = E$.

10. **False.** If $B = C = D = E$ then $\rho(A) = m$.

Exercises

1.
$$E_{(1),(3)}A = \begin{bmatrix} 1 & -3 & -1 \\ 3 & 1 & 1 \\ 2 & 4 & 2 \end{bmatrix}; \quad E_{(2) - 1(1)}E_{(1),(3)}A = \begin{bmatrix} 1 & -3 & -1 \\ 2 & 4 & 2 \\ 2 & 4 & 2 \end{bmatrix};$$

$$E_{(3) - 1(2)}E_{(2) - 1(1)}E_{(1),(3)}A = \begin{bmatrix} 1 & -3 & -1 \\ 2 & 4 & 2 \\ 0 & 0 & 0 \end{bmatrix};$$

$$E_{(2) - 2(1)}E_{(3) - 1(2)}E_{(2) - 1(1)}E_{(1),(3)}A = \begin{bmatrix} 1 & -3 & -1 \\ 0 & 10 & 4 \\ 0 & 0 & 0 \end{bmatrix};$$

$$E_{\frac{1}{10}(2)}E_{(2) - 2(1)}E_{(3) - 1(2)}E_{(2) - 1(1)}E_{(1),(3)}A = \begin{bmatrix} 1 & -3 & -1 \\ 0 & 1 & \frac{2}{5} \\ 0 & 0 & 0 \end{bmatrix}.$$

2.
$$E = E_{\frac{1}{10}(2)}E_{(2)-2(1)}E_{(3)-1(2)}E_{(2)-1(1)}E_{(1),(3)}$$

$$= \begin{bmatrix} 0 & 0 & 1 \\ 0 & \frac{1}{10} & -\frac{3}{10} \\ 1 & -1 & 1 \end{bmatrix}.$$

$$E^{-1} = \begin{bmatrix} 2 & 10 & 1 \\ 3 & 10 & 0 \\ 1 & 0 & 0 \end{bmatrix}.$$

3. Let $E = E_{(1)-1(2)}E_{-\frac{1}{2}(2)}E_{(2)+1(3)}E_{(1)-1(3)}E_{\frac{1}{2}(3)}E_{(3)+1(1)}E_{(2)-3(1)}$. Then $EC = I_3$. Therefore

$$C^{-1} = E = \begin{bmatrix} -\frac{3}{4} & \frac{1}{2} & -\frac{1}{4} \\ \frac{5}{4} & -\frac{1}{2} & -\frac{1}{4} \\ \frac{1}{2} & 0 & \frac{1}{2} \end{bmatrix}.$$

4. If $A_i \in M_{n_i}(R)$, $i = 1, \ldots, k$, and $\sum_{i=1}^{k} n_i = n$, then, by Theorem 2.5,

$$\left(\sum_{i=1}^{k} A_i \right) \left(\sum_{i=1}^{k} A_i^{-1} \right) = \sum_{i=1}^{k} A_i A_i^{-1} = \sum_{i=1}^{k} I_{n_i} = I_n.$$

5. Let U be the subspace consisting of all n-tuples $u = (u_1, \ldots, u_n)$ with $u_j = 0, j = k+1, \ldots, n$. Then

$$Tu = (B + C)u = \left(\sum_{j=1}^{k} b_{1j}u_j, \ldots, \sum_{j=1}^{k} b_{kj}u_j, \right.$$
$$\left. \sum_{j=k+1}^{n} c_{1j}u_j, \ldots, \sum_{j=k+1}^{n} c_{n-k,j}u_j \right)$$
$$= \left(\sum_{j=1}^{k} b_{1j}u_j, \ldots, \sum_{j=1}^{k} b_{kj}u_j, 0, \ldots, 0 \right) \in U.$$

$U^{\perp}$ clearly consists of all n-tuples $v = (v_1, \ldots, v_n)$ with $v_j = 0, j = 1, \ldots, k$. We have for $v \in U^{\perp}$

$$Tv = (B + C)v = \left(\sum_{j=1}^{k} b_{1j}v_j, \ldots, \sum_{j=1}^{k} b_{kj}v_j \sum_{j=k+1}^{n} c_{1j}v_j, \ldots, \right.$$
$$\left. \sum_{j=k+1}^{n} c_{n-k,j}u_j \right)$$
$$= \left(0, \ldots, 0, \sum_{j=k+1}^{n} c_{1j}v_j, \ldots, \sum_{j=k+1}^{n} c_{n-k,j}u_j \right) \in U^{\perp}.$$

Section 2.3

Quiz

1. **False.** A system $Ax = b$ has no solutions if $b \notin \langle A^{(1)}, \ldots, A^{(n)} \rangle$. [See Theorem 3.3 (c).] For example, the system of 2 equations in 3 unknowns,

$$x_1 + x_2 + x_3 = 0$$
$$x_1 + x_2 + x_3 = 1,$$

    has no solutions.

2. **False.** If $b \neq 0$ then the zero vector is not a solution vector of $Ax = b$.

3. **True.** This follows immediately from Theorems 3.2 (b) and 3.4 (b).

4. **True.** In the notation of Theorem 3.1, if $i > r$ then $a_{ij} = 0$ for all j. If $i = 1, \ldots, r$, then $a_{ij} = 0$ for all $j < n_i$ and, since $i \leq n_i$, $i = 1, \ldots, r$, we have $a_{ij} = 0$ for all $j < i$.

5. **True.** If A^{-1} exists then A is nonsingular and, by Theorem 3.2 (a), the Hermite normal forms of both A^{-1} and A are equal to the identity matrix.

6. **True.** If $\rho(A) = m$ then the columns of A span $V_m(R)$ and therefore $b \in C(A)$ and, by Theorem 3.3 (c), the system $Ax = b$ has a solution.

7. **True.** The matrix A is row equivalent to its normal Hermite form. By Theorem 2.2, a matrix is equivalent to a zero matrix if and only if it is equal to it.

8. **True.** If A^{-1} and B^{-1} exist then they are nonsingular and therefore, by Theorem 3.2 (a), both have I_n as their Hermite normal form.

9. **True.** For, if v is a solution of $Ax = b$, then $Av = b$ and thus $v = A^{-1}b$.

10. **False.** The condition is certainly sufficient. For, if we denote the $m \times n$ matrix (a_{ij}) by A, then $Ax = 0$ has nonzero solutions if and only if $\rho(A) < n$ [see Theorem 3.3 (a)]. Now, $\rho(A) \leq \min (m,n)$. The condition is obviously not necessary. For example, the system $2^i x_1 + 2^i x_2 = 0$, $i = 1,2,3$, has a nonzero solution: $x_1 = -x_2 = 1$.

Exercises

1. Let $A \in M_{m,n}(R)$, $B \in M_{m,n}(R)$ and $\rho(A) = \rho(B) = r$. Let $C = (c_{ij})$ and $D = (d_{ij})$ be the canonical forms of A and B respectively. Then $c_{ii} = d_{ii} = 1$, $i = 1, \ldots, r$, and $c_{ij} = d_{ij} = 0$ otherwise. Hence $C = D$. Now, by Theorem 3.4 (a), A is equivalent to C and B is equivalent to D. Thus A and B are equivalent. Conversely, if A and B are equivalent, then A can be obtained from B by elementary row and column operations. It follows from Theorems 2.2 and 2.3 (b) that $\rho(A) = \rho(B)$.

2. $$\mathrm{I}_{(i),(j)} = \mathrm{III}_{-1(i)} \mathrm{II}_{(j)+1(i)} \mathrm{II}_{(i)-1(j)} \mathrm{II}_{(j)+1(i)}.$$

3. Let $A \in M_n(R)$, $\rho(A) = r < n$ [see Theorem 3.2 (b)]. Then, by Theorem

3.4 (a), there exist nonsingular n-square matrices P and Q such that

$$PAQ = I_r \dotplus 0_{n-r}$$

Let $B = Q(0_r \dotplus I_{n-r})$. Then $B \neq 0$ and $PAB = (I_r \dotplus 0_{n-r})(0_r \dotplus I_{n-r}) = 0_n$. Since P is nonsingular, we have

$$AB = 0_n.$$

4. Denote the system by

(i) $$Ax = b$$

and proceed as in the example following Theorem 3.1. Perform on A elementary row operations $I_{(1),(2)}, II_{(2)-2(1)}, II_{(3)-2(1)}, II_{(4)-2(1)}, III_{-1(2)},$ $II_{(1)-1(2)}, II_{(3)+1(2)}, II_{(1)+1(3)}, II_{(2)-2(3)}, II_{(4)+1(3)}$ in order and obtain

(ii) $$EAx = Eb$$

where $E = E_{(4)+1(3)}E_{(2)-2(3)}E_{(1)+1(3)}E_{(3)+1(2)}E_{(1)-1(2)}E_{-1(2)}E_{(4)-2(1)}$ $E_{(3)-2(1)}E_{(2)-2(1)}E_{(1),(2)}$

$$= \begin{bmatrix} 0 & -1 & 1 & 0 \\ 1 & 2 & -2 & 0 \\ -1 & 0 & 1 & 0 \\ -1 & -2 & 1 & 1 \end{bmatrix}.$$

Thus (ii) becomes

$$\begin{bmatrix} 1 & 0 & 0 & 1 & 0 \\ 0 & 1 & 0 & -1 & 3 \\ 0 & 0 & 1 & 1 & -2 \\ 0 & 0 & 0 & 0 & 0 \end{bmatrix} x = \begin{bmatrix} 0 \\ 3 \\ 0 \\ 0 \end{bmatrix}.$$

Now, using the notation introduced immediately after the proof of Theorem 3.2, $N = \{1,2,3\}$ and

$$u_4 = e_4 - e_1 + e_2 - e_3 = (-1,1,-1,1,0),$$
$$u_5 = e_5 - 3e_2 + 2e_3 = (0,-3,2,0,1),$$
$$c' = 3e_2 = (0,3,0,0,0).$$

Therefore every solution x of (i) is of the form

$$x = c' + x_4 u_4 + x_5 u_5.$$

5. We use the method used in the proof of Theorem 3.2. We find that

$$A = E_{(1),(4)}E_{(2)+1(1)}E_{(3)+1(1)}E_{(4)+2(1)}E_{(1)+1(2)}E_{(4)-1(2)}E_{(1)+1(3)}$$
$$E_{(4)-1(3)}E_{-5(4)}E_{(1)+4(4)}E_{(2)-1(4)}E_{(3)-1(4)}.$$

Hence

$$A^{-1} = E_{(3)+1(4)}E_{(2)+1(4)}E_{(1)-4(4)}E_{-\frac{1}{5}(4)}E_{(4)+1(3)}E_{(1)-1(3)}E_{(4)+1(2)}$$
$$E_{(1)-1(2)}E_{(4)-2(1)}E_{(3)-1(1)}E_{(2)-1(1)}E_{(1),(4)}$$

$$= \frac{1}{5}\begin{bmatrix} 4 & -1 & -1 & -1 \\ -1 & 4 & -1 & -1 \\ -1 & -1 & 4 & -1 \\ -1 & -1 & -1 & 4 \end{bmatrix}.$$

Section 2.4

Quiz

1. **False.** For example, $N(\sigma\sigma^{-1}) = 0$ for any n.
2. **False.** By Theorem 4.3, $\epsilon(\sigma_m \cdots \sigma_1) = (-1)^m$ and, for odd n, $(-1)^m \neq (-1)^{n-m}$.
3. **False.** For example, $(1234)^2 = (13)(24)$.
4. **True.** Note that $(i_1i_2) = (i_2i_1)$ and that the two transpositions cannot be disjoint, for then they would commute. Thus either $i_1 = j_1$ and $j_2 < i_2$, or $i_1 < j_1$ and $j_2 = i_2$, or $i_1 = j_1$ and $j_2 = i_2$. If $i_1 = j_1$ and $j_2 < i_2$, then $(i_1i_2)(j_1j_2) = (i_1i_2)(i_1j_2) = (i_1j_2i_2)$, while $(j_1j_2)(i_2i_1) = (i_1i_2j_2)$. Similarly in case $i_1 < j_1$ and $j_2 = i_2$. On the other hand, if $i_1 = j_1$ and $j_2 = i_2$, then $(i_1i_2)(j_1j_2)$ and $(j_1j_2)(i_2i_1)$ are both equal to the identity permutation.
5. **True.** Any 1-cycle in S_n is equal to the identity permutation.
6. **True.** If σ is the k-cycle $(i_1, \sigma(i_1), \ldots, \sigma^{k-1}(i_1))$ then $\sigma^k(i_1) = i_1$, by Def. 4.3. Therefore $\sigma^k(\sigma^t(i_1)) = \sigma^t(\sigma^k(i_1)) = \sigma^t(i_1)$, $t = 1, \ldots, k-1$.
7. **False.** $(12)(13) = (132)$.
8. **False.** $(1234)^{-1} = (4321)$.
9. **True.** $\tau\sigma\tau^{-1} = (3412) = (1234)$ (see Exercise 5).
10. **False.** $\varphi\sigma\varphi^{-1} = (4321) \neq (1234)$.

Exercises

1. Any even permutation is a product of an even number of transpositions. We prove that a product of any two transpositions is a product of 3-cycles and the result will follow. If i_1,i_2,i_3,i_4 are distinct, then $(i_1i_2)(i_3i_4) = (i_1i_2i_3)(i_2i_3i_4)$. If the two transpositions have a number in common, then (i_1, i_2 and i_3 are assumed to be distinct) $(i_1i_2)(i_1i_3) = (i_1i_3i_2)$. Finally, if the two transpositions are equal, then their product is equal to the identity permutation and we can write $(i_1i_2)(i_1i_2) = e = (123)(132)$.
2. (a) $\sigma = (13278)(46)(5)$
    (b) $\sigma = (18)(17)(12)(13)(46)$.

3. $N(\sigma) = (5 - 1) + (2 - 1) + (1 - 1) = 5$. Therefore $\epsilon(\sigma) = (-1)^5 = -1$. On the other hand, the number of transpositions in Question 2 (b) is 5 and thus

$$\epsilon(\sigma) = (-1)^5 = -1.$$

4. If each integer $1, \ldots, n$ appears in exactly one of $\sigma_1, \ldots, \sigma_m$, then $n = 2m$ and therefore

$$\epsilon(\sigma) = (-1)^{n-m} = (-1)^{2m-m} = (-1)^m.$$

If some integers do not appear in any of $\sigma_1, \ldots, \sigma_m$, then $N(\sigma) < n - m$. For example, if $\sigma = (12)(34) \in S_5$, then $N(\sigma) = 2 < 5 - 2$.
5. For, $\tau\sigma\tau^{-1}(\tau(i_j)) = \tau\sigma(i_j) = \tau(i_{j+1})$, $j = 1, \ldots, k$, where $i_{k+1} = i_1$.
6. Let $\sigma_1, \ldots, \sigma_k$ be all the distinct even permutations of S_n. Consider the permutations $(12) \sigma_1, \ldots, (12) \sigma_k$. They are all distinct and odd. Thus the number of even permutations of S_n cannot exceed the number of odd permutations of S_n. By reversing the roles of even and odd permutations in the above argument, we see that the number of odd permutations of S_n cannot exceed the number of even permutations of S_n. Thus the two are equal and $k = n!/2$.
7. A sequence of k distinct integers $i_1, \ldots, i_k$ can be chosen from $1, \ldots, n$ in $n!/(n - k)!$ ways. To the sequence $i_1, \ldots, i_k$ make correspond the k-cycle $(i_1 \cdots i_k)$. For each such sequence there are $k - 1$ other sequences obtained by cyclically permuting $i_1, \ldots, i_k$, and these correspond to the same k-cycle: i.e., $(i_1 \cdots i_k) = (i_2 \cdots i_k i_1) = \cdots = (i_k i_1 \cdots i_{k-1})$. Thus the total number of k-cycles in S_n is $n!/(n - k)!k$.

Section 2.5

Quiz

1. **False.** Take $n = 3$, $A = \begin{bmatrix} 0 & 0 & 1 \\ 0 & 0 & 1 \\ 1 & 1 & 1 \end{bmatrix}$.
2. **False.** Take $n = 2$, $A = \begin{bmatrix} 1 & 1 \\ 1 & 1 \end{bmatrix}$. Then $A^2 = 2A$, per $(A^2) = 4$ per (A).
3. **False.** The example in the preceding exercise will do.
4. **True.** From Theorem 5.2, $\det(e_{\varphi(1)}, \ldots, e_{\varphi(n)}) = \epsilon(\varphi) \det(e_1, \ldots, e_n) = \epsilon(\varphi)$.
5. **False.** For, per $(I_2) = \det(I_2) = 1$.
6. **True.** For α' is just the ordered complementary set and hence is in $Q_{n-r,n}$.

7. **False.** In fact, $A[1,1|3,4] = \begin{bmatrix} a_{13} & a_{14} \\ a_{13} & a_{14} \end{bmatrix}$.

8. **True.**

9. **True.**

10. **True.** By definition, to include rows α' means to exclude rows α.

Exercises

1. By definition, per $(J_n) = \sum_{\sigma \in S_n} \left(\frac{1}{n}\right)^n = \frac{n!}{n^n}$.

   On the other hand, the rank of J_n is 1, J_n is singular [Theorem 3.2 (b)] and hence by Theorem 5.3, det $(J_n) = 0$.

2. If two rows or columns of A are the same, then the rank of A is less than n and, by Theorem 3.2 (b), A is singular. Hence, by Theorem 5.3, det $(A) = 0$.

3. The coefficient matrix is

$$A = \begin{bmatrix} 1 & -1 & 0 & 0 \\ 0 & 1 & 1 & 0 \\ 0 & 0 & 1 & 1 \\ 1 & 0 & 0 & 1 \end{bmatrix},$$

and det $(A) = 2$. Thus A is nonsingular and hence $Ax = 0$ implies $x = A^{-1}Ax = 0$.

4. Let $B = A^T$. Then per $(B) = \sum_{\sigma \in S_n} \prod_{i=1}^{n} b_{i\sigma(i)}$. But $b_{ij} = a_{ji}$, hence $b_{i\sigma(i)} = a_{\sigma(i)i}$. It follows that

$$\text{per } (A^T) = \sum_{\sigma \in S_n} \prod_{i=1}^{n} a_{\sigma(i)i}.$$

Set $j = \sigma(i)$ in $\prod_{i=1}^{n} a_{\sigma(i)i}$ to obtain $\prod_{j=1}^{n} a_{j\sigma^{-1}(j)}$. But σ^{-1} runs over S_n precisely once as σ does. Hence

$$\sum_{\sigma \in S_n} \prod_{i=1}^{n} a_{\sigma(i)i} = \sum_{\sigma^{-1} \in S_n} \prod_{j=1}^{n} a_{j\sigma^{-1}(j)}$$

$$= \sum_{\varphi \in S_n} \prod_{i=1}^{n} a_{i\varphi(i)} = \text{per } (A).$$

5.   As we saw in the example preceding Def. 5.3, this number is just

$$
\text{per}
\begin{bmatrix}
0 & 1 & 1 & 1 \\
1 & 0 & 1 & 1 \\
1 & 1 & 0 & 1 \\
1 & 1 & 1 & 0
\end{bmatrix}.
$$

A straightforward computation yields 9 for this value.

6.   We compute

$$
|\text{per}\,(A)| = \left| \sum_{\sigma \in S_n} \prod_{i=1}^{n} a_{i\sigma(i)} \right| \leq \sum_{\sigma \in S_n} \prod_{i=1}^{n} |a_{i\sigma(i)}|
$$

$$
= \sum_{\sigma \in S_n} \prod_{i=1}^{n} b_{i\sigma(i)} = \text{per}\,(B).
$$

7.   A sequence $\alpha = (\alpha_1, \ldots, \alpha_r)$ in $\Gamma_{r,n}$ can be constructed by choosing α_i to be any one of $1, \ldots, n$ for each i. Thus there are n^r possible choices for α.

8.   The elements of a sequence $\alpha \in Q_{r,n}$ may be chosen by just selecting a set of r different numbers from $1, \ldots, n$ and putting them in increasing size. There are $\binom{n}{r}$ selections possible from n items.

9.   Let

$$
\beta \in Q_{r,n+r-1}, \quad \beta = (\beta_1, \beta_2, \ldots, \beta_r).
$$

Then $(\beta_1, \beta_2 - 1, \beta_3 - 2, \ldots, \beta_r - (r-1))$ is a sequence in $G_{r,n}$. Conversely, if $(\alpha_1, \ldots, \alpha_r)$ is a sequence in $G_{r,n}$ then $\alpha_1 < \alpha_2 + 1 < \alpha_3 + 2 < \cdots < \alpha_r + (r-1)$ so that $\beta = (\alpha_1, \alpha_2 + 1, \ldots, \alpha_r + (r-1))$ is in $Q_{r,n+r-1}$. Hence there is a one-one correspondence between the sequences in $Q_{r,n+r-1}$ and the sequences in $G_{r,n}$. It follows from Exercise 8 that there are $\binom{n+r-1}{r}$ sequences in $Q_{r,n+r-1}$ and hence this number of sequences in $G_{r,n}$.

10.   Let A be an $m \times n$ matrix. According to Exercise 8, there are $\binom{m}{r}$ selections of α and $\binom{n}{s}$ selections for β in $A[\alpha|\beta]$. Thus the number of $r \times s$ submatrices of A is $\binom{m}{r}\binom{n}{s}$.

Section 2.6

Quiz

1.   **True.** For, $\det\,(AA^*) = \det\,(A)\det\,(A^*) = \det\,(A)\,\overline{\det\,(A)} = |\det\,(A)|^2 \geq 0.$

2.   **False.** Let $A = \begin{bmatrix} 1 & 2 \\ 2 & 1 \end{bmatrix}$. Then $\det\,(A) = -3$.

3. **True.** If we expand det (B) using rows $1, \ldots, n$ by the Laplace expansion theorem then

$$\det (B) = (-1)^{\frac{n(n+1)}{2}} \sum_{\omega \in Q_{n,2n}} (-1)^{s(\omega)} \det (B[1, \ldots, n|\omega])$$
$$\det (B(1, \ldots, n|\omega)).$$

Now det $(B[1, \ldots, n|\omega]) = 0$ unless $\omega = \omega_0 = (n+1, \ldots, 2n)$. Hence

$$\det (B) = (-1)^{\frac{n(n+1)}{2}} (-1)^{s(\omega_0)} \det (B[1, \ldots, n|\omega_0])$$
$$\det (B(1, \ldots, n|\omega_0)).$$

Now det $(B[1, \ldots, n|\omega_0]) = \det (A)$, $\det (B(1, \ldots, n|\omega_0)) = \det (A^*)$. Hence

$$\det (B) = (-1)^{\frac{n(n+1)}{2} + s(\omega_0)}|\det (A)|^2 = (-1)^{2n(2n+1)/2}|\det (A)|^2$$
$$= (-1)^n|\det (A)|^2.$$

4. **True.** Theorem 6.8.
5. **True.** Expanding det (A) by the Laplace expansion on the first two columns we have

$$\det (A) = (-1)^{(1+2)+(3+4)} \det (A[3,4|1,2]) \det (A(3,4|1,2))$$

$$= \det \begin{bmatrix} 2 & 3 \\ 6 & 7 \end{bmatrix} \det \begin{bmatrix} 1 & 0 \\ 0 & -1 \end{bmatrix} = -4(-1) = 4.$$

6. **True.** For, every term in the Laplace expansion of A using rows α must be 0.
7. **False.** Take $n = 2$, $k = 1$, $A = \begin{bmatrix} 1 & 1 \\ 1 & 1 \end{bmatrix}$. Then A is singular although $A[i|j] = A(i|j) = [1]$ are nonsingular.
8. **True.** Let A_1 and A_2 be p and q square matrices respectively. Then $A_1 \dotplus A_2 = (A_1 \dotplus I_q)(I_p \dotplus A_2)$. By Theorem 6.2, det $(A_1 \dotplus A_2) = \det (A_1 \dotplus I_q) \det (I_p \dotplus A_2) = \det (A_1) \det (A_2)$. The case for more than two matrices is done by a trivial induction.
9. **True.** We compute $(\operatorname{adj} A)^{(3)} = ((-1)^4 \det (A(3|1)), (-1)^5 \det (A(3|2)), (-1)^6 \det (A(3|3))) = (7, -5, 1)$.
10. **True.** For, $UU^* = I_n$ and hence $|\det (U)|^2 = 1$.

Exercises

1. Let $k > \min \{\rho(A), \rho(B)\}$. Then by Theorem 6.6, $\det (A[\alpha|\beta]) \det (B[\beta|\gamma])$
    $= 0$ for any α and β in $Q_{k,n}$. Hence, by Theorem 6.1,

    $$\det (AB[\alpha|\gamma]) = \sum_{\beta \in Q_{k,n}} \det (A[\alpha|\beta]) \det (B[\beta|\gamma]) = 0.$$

    Theorem 6.6 then implies that $\rho(AB) < k$.

2. If $Ax = 0$ then $A^*Ax = 0$. Conversely, if $A^*Ax = 0$ then $0 = (A^*Ax,x)$
    $= (Ax,Ax) = \|Ax\|^2$ so $Ax = 0$. Thus $\rho(A) = \rho(AA^*)$. The argument
    used in Theorem 6.8 can also be used here.

3. If A is a unit, then $1 = \det (I_n) = \det (AA^{-1}) = \det (A) \det (A^{-1})$.
    Hence $\det (A^{-1})$ must be ±1. Conversely, if $\det (A^{-1}) = \pm1$, then
    $\det (A) = \pm1$ and hence $A^{-1} = \dfrac{1}{\det (A)}$ adj A has integer entries.

4. If A and B are unit matrices with integer entries, then $\det (AB) =$
    $\det (A) \det (B) = \pm1$. Use Exercise 3.

5. Let α and ω be fixed. A typical term in per $(A[\alpha|\omega])$ is $a_{\alpha_1\omega_{\sigma(1)}} \cdots a_{\alpha_k\omega_{\sigma(k)}}$,
    $\sigma \in S_k$. A typical term in per $(A(\alpha|\omega))$ is $a_{\alpha'_1\omega'_{\varphi(1)}} \cdots a_{\alpha'_{n-k}\omega'_{\varphi(n-k)}}$ where
    $\varphi \in S_{n-k}$. Then a typical term in the product per $(A[\alpha|\omega])$ per $(A(\alpha|\omega))$
    is just

    $$a_{\alpha_1\omega_{\sigma(1)}} \cdots a_{\alpha_k\omega_{\sigma(k)}}a_{\alpha'_1\omega'_{\varphi(1)}} \cdots a_{\alpha'_{n-k}\omega'_{\varphi(n-k)}} = a_{1\theta(1)} \cdots a_{n\theta(n)},$$

    where $\theta \in S_n$ and is, in fact, the uniquely defined permutation

    $$\theta = \begin{pmatrix} \alpha_1 & \cdots & \alpha_k & \alpha'_1 & \cdots & \alpha'_{n-k} \\ \omega_{\sigma(1)} & \cdots & \omega_{\sigma(k)} & \omega'_{\varphi(1)} & \cdots & \omega'_{\varphi(n-k)} \end{pmatrix}.$$

    Every term in per (A) appears in precisely one of the products
    per $(A[\alpha|\omega])$ per $(A(\alpha|\omega))$.

6. Suppose first that $m \leq n$. Then

    $$\det (AA^T) = \sum_{\omega \in Q_{m,n}} \det (A[1, \ldots, m|\omega]) \det (A^T[\omega|1, \ldots, m]).$$

    But $\det (A^T[\omega|1, \ldots, m]) = \det (A[1, \ldots, m|\omega])$. Thus $\det (AA^T) =$
    $\sum_{\omega \in Q_{m,n}} (\det (A[1, \ldots, m|\omega]))^2 \geq 0$. If $m > n$ then $m > n \geq \rho(A) =$
    $\rho(AA^T)$. Hence AA^T is singular and $\det (AA^T) = 0$.

7. If $\beta = \alpha$ this is just the Laplace expansion theorem. Suppose $\alpha \neq \beta$.
    Then $\beta' \cap \alpha$ is not empty. That is, there are integers in α and not in β.
    Now define a matrix C as follows: $C_{(\alpha_1)} = A_{(\alpha_1)}$, $C_{(\alpha_2)} = A_{(\alpha_2)}$, $\ldots$,
    $C_{(\alpha_k)} = A_{(\alpha_k)}$, $C_{(\alpha'_1)} = A_{(\beta'_1)}$, $\ldots$, $C_{(\alpha'_{n-k})} = A_{(\beta'_{n-k})}$. Since $\beta' \cap \alpha$ is not

empty, two rows of C (at least) must be the same. Thus det $(C) = 0$. It is also obvious that $C[\alpha|\omega] = A[\alpha|\omega]$ and $C(\alpha|\omega) = C[\alpha'|\omega'] = A[\beta'|\omega'] = A(\beta|\omega)$. Hence det $(C[\alpha|\omega]) = $ det $(A[\alpha|\omega])$, det $(C(\alpha|\omega)) = $ det $(A(\beta|\omega))$. Thus

$$
\begin{aligned}
0 = \det (C) &= \sum_{\omega \in Q_{k,n}} (-1)^{s(\alpha) + s(\omega)} \det (C[\alpha|\omega]) \det (C(\alpha|\omega)) \\
&= \sum_{\omega \in Q_{k,n}} (-1)^{s(\alpha) + s(\omega)} \det (A[\alpha|\omega]) \det (A(\alpha|\omega)).
\end{aligned}
$$

CHAPTER **3**

Section 3.1

Quiz

1. **False.** The matrices $\begin{bmatrix} 0 & 0 \\ 0 & 0 \end{bmatrix}$ and $\begin{bmatrix} 0 & 1 \\ 0 & 0 \end{bmatrix}$ have the same characteristic roots but they are not similar.

2. **False.** According to Theorem 1.5 (c), when $r = 1$, similar matrices have the same trace. But these two matrices do not.

3. **True.** By Theorem 1.4, det $(\lambda I_2 - A) = \lambda^2 - E_1(A)\lambda + E_2(A)$. However, $E_1(A) = $ tr (A) and $E_2(A) = $ det (A) for 2-square matrices.

4. **False.** $(\sqrt[3]{2})^3 - 2 = 0$ and thus $\sqrt[3]{2}$ is a root of a polynomial with rational coefficients.

5. **True.** They both have the characteristic polynomial $(\lambda - 1)(\lambda - 2)(\lambda - 3)$.

6. **False.** If $c = 0$ then $cx = 0$ and characteristic vectors are non-zero by definition.

7. **True.** If $A = 0$ then clearly $AA^* = 0$, so $\mu_1 + \cdots + \mu_n = 0$. Conversely, if $\mu_1 + \cdots + \mu_n = 0$ then tr $(AA^*) = 0$. By Theorem 1.6, tr $(AA^*) = \sum_{i,j=1}^{n} |a_{ij}|^2$. Hence $A = 0$.

8. **True.** For, $De_i = d_i e_i$, $i = 1, \ldots, n$.

9. **True.** For, det $(A) = \prod_{i=1}^{n} \lambda_i$ from Theorem 1.5 (a).

10. **True.** We know from Theorem 5.3, Chap. 2, that T is singular if and only if $[T]_G{}^G = A$ is singular for any basis G of V. But the characteristic roots of T and A are the same and A is singular if and only if det $(A) = 0$ (Chap. 2, Theorem 6.4). Reference to the preceding question completes the argument.

Exercises

1. Suppose $Tu = ru$ and $Su \neq 0$. Then $T(Su) = (TS)u = (ST)u = S(Tu)$ $= S(ru) = rSu$. Thus Su is a characteristic vector of T corresponding to r.

2. Observe that $STu = S\mu_i u = \mu_i \lambda_i u$ so that $\mu_i \lambda_i$ is a characteristic root of ST. Similarly, it is a characteristic root of TS. It follows by an easy induction on k that if $k \geq 1$, then λ_i^k is a characteristic root of S^k corresponding to u.

3. This is trivial to see: $(cT)u = cTu = c\lambda_j u$ where u is a characteristic vector of T corresponding to λ_j.

4. Consider the characteristic polynomial of $cI_n + A$:

$$\det [\lambda I_n - (cI_n + A)] = \det [(\lambda - c)I_n - A].$$

Now, if we let $p(\lambda)$ denote the characteristic polynomial of A then the characteristic polynomial of $cI_n + A$ is just $p(\lambda - c)$. Hence, the characteristic roots of $cI_n + A$ are the values $c + \lambda_i$, $i = 1, \ldots, n$.

5. Suppose $Tv = rv$, $v \neq 0$. Then $v_1 - v_2 = rv_1$, $v_1 + v_2 = rv_2$. It follows that $(1 - r)v_1 = v_2$, $(1 - r)v_2 = -v_1$. Neither v_1 nor v_2 can be 0 otherwise $v = 0$. Hence

$$1 - r = \frac{v_2}{v_1} = -\frac{1}{1 - r}, \quad (1 - r)^2 = -1, \quad (1 - r) = \pm i, \quad r = 1 \pm i.$$

By replacing r by $1 + i$ and then by $1 - i$ in $v_1 - v_2 = rv_1$, $v_1 + v_2 = rv_2$ we can check that there are no real nonzero vectors v satisfying $Tv = rv$.

6. The characteristic polynomial is directly computed to be

$$\det (\lambda I_n - A) = \lambda^3 - 2\lambda^2 - \lambda + 2 = (\lambda - 1)(\lambda - 2)(\lambda + 1).$$

Thus the characteristic roots are $r_1 = 1$, $r_2 = -1$, $r_3 = 2$. One computes immediately that $(1,1,0)$ is a characteristic vector corresponding to 1, $(1,1,1)$ corresponds to -1, and $(1,0,1)$ corresponds to 2. For example, the system $Ax = r_1 x$ becomes

$$3x_1 - 3x_2 - 2x_3 = 0$$
$$2x_1 - 2x_2 - 2x_3 = 0$$
$$3x_1 - 3x_2 - 2x_3 = 0.$$

This system is equivalent to

$$x_1 - x_2 \qquad = 0$$
$$x_1 - x_2 - x_3 = 0.$$

Thus $(x_1,x_2,x_3) = (x_1,x_1,0) = x_1(1,1,0)$. In other words, any characteristic vector of A corresponding to $r_1 = 1$ is a multiple of $(1,1,0)$. Also, $E_3(A) =$ det $(A) = -2$, $E_2(A) = \sum_{\omega \in Q_{2,3}}$ det $(A[\omega|\omega]) = -1$, $E_1(A) = $ tr $(A) = 2$. Moreover, $E_1(r_1,r_2,r_3) = 2$, $E_2(r_1,r_2,r_3) = -1$, $E_3(r_1,r_2,r_3) = -2$.

7. We compute that $E_3(B) = 4$, $E_2(B) = 8$, $E_1(B) = 5$. The characteristic polynomial is then $\lambda^3 - 5\lambda^2 + 8\lambda - 4 = (\lambda - 1)(\lambda - 2)^2$. A characteristic vector corresponding to 1 is $(1,1,1)$ and characteristic vectors corresponding to 2 are $(1,2,1)$ and $(1,1,0)$.

8. We compute that $E_3(C) = 1$, $E_2(C) = 3$, $E_1(C) = 3$. Thus the characteristic polynomial of C is $\lambda^3 - 3\lambda^2 + 3\lambda - 1 = (\lambda - 1)^3$. Hence 1 is a characteristic root of multiplicity 3. The equation $Ax = x$ becomes

$$3x_1 + 2x_2 - 6x_3 = x_1$$
$$2x_2 - x_3 = x_2$$
$$x_1 + x_2 - 2x_3 = x_3.$$

    This system of equations has the solutions $(x_1,x_2,x_3) = c(2,1,1)$ where c is any scalar.

9. The characteristic polynomial of A^T is det $(\lambda I_n - A^T) = $ det $((\lambda I_n - A)^T) = $ det $(\lambda I_n - A)$. Hence A^T and A have the same characteristic polynomial. If $c = 0$ the second assertion is trivial. If $c \neq 0$ then

$$\text{det } (\lambda I_n - cA) = \text{det } \left(c \left(\frac{\lambda}{c} I_n - A \right) \right) = c^n \text{ det } \left(\frac{\lambda}{c} I_n - A \right). \text{ Thus } r \text{ is}$$

    a characteristic root of cA if and only if $\lambda_i = \dfrac{r}{c}$ for some characteristic root λ_i of A.

Section 3.2

Quiz

1. **False.**   The characteristic roots of $A = \begin{bmatrix} 2 & 0 \\ 3 & 1 \end{bmatrix}$ are 2, 1 (see Theorem 2.5) and those of $B = \begin{bmatrix} 3 & 3 \\ 0 & 4 \end{bmatrix}$ are 3, 4. The characteristic roots of $A + B = \begin{bmatrix} 5 & 3 \\ 3 & 5 \end{bmatrix}$ are 8, 2.

2. **True.**   Each of these matrices has three distinct characteristic roots 1, 2, 3 and, by Theorem 2.3, is similar to the diagonal matrix diag $(1,2,3)$.

3. **True.**   By Theorem 2.3, there exists a matrix $P \in M_2(R)$ such that

$$P^{-1} \begin{bmatrix} 2 & 1 \\ 0 & 1 \end{bmatrix} P = \begin{bmatrix} 2 & 0 \\ 0 & 1 \end{bmatrix}$$

    $\left(\text{as a matter of fact, } P = \begin{bmatrix} 1 & -1 \\ 0 & 1 \end{bmatrix} \right).$

Let $S = I_1 \dotplus P$. Then $S^{-1} = I_1 \dotplus P^{-1}$ and

$$S^{-1} \begin{bmatrix} 2 & 0 & 0 \\ 0 & 2 & 1 \\ 0 & 0 & 1 \end{bmatrix} S = \begin{bmatrix} 2 & 0 & 0 \\ 0 & 2 & 0 \\ 0 & 0 & 1 \end{bmatrix}.$$

4. **False.** Both roots of the nonzero matrix $\begin{bmatrix} 0 & 1 \\ 0 & 0 \end{bmatrix}$ are 0.

5. **True.** Let $P = (p_{ij})$ be the n-square permutation matrix with $p_{ij} = \delta_{i\sigma(j)}$. Then

$$(AP)_{tj} = \sum_{s=1}^{n} a_{ts}\delta_{s\sigma(j)} = a_{t\sigma(j)} = d_t\delta_{t\sigma(j)}$$

and

$$\begin{aligned} (P^{-1}AP)_{ij} &= \sum_{t=1}^{n} (P^{-1})_{it}(AP)_{tj} \\ &= \sum_{t=1}^{n} \delta_{i\sigma^{-1}(t)}\delta_{t\sigma(j)}d_t \\ &= \delta_{ij}d_{\sigma(j)}. \end{aligned}$$

6. **True.** If $(P)_{ij} = \delta_{i\sigma(j)}$, then $(P^T)_{ij} = \delta_{j\sigma(i)}$. Thus

$$(PP^T)_{ij} = \sum_{t=1}^{n} \delta_{i\sigma(t)}\delta_{j\sigma(t)} = \delta_{ij}$$

and therefore $PP^T = I_n$. Hence $P^T = P^{-1}$. The result follows from the definition of similarity and the solution of the preceding question.

7. **True.** We proved in the preceding solution that $PP^T = I_n$.

8. **False.** Let $P = \begin{bmatrix} 0 & 1 & 0 \\ 0 & 0 & 1 \\ 1 & 0 & 0 \end{bmatrix}$, $Q = \begin{bmatrix} 0 & 0 & 1 \\ 1 & 0 & 0 \\ 0 & 1 & 0 \end{bmatrix}$ and $S = \begin{bmatrix} 1 & 0 & 0 \\ 0 & 0 & 1 \\ 0 & 1 & 0 \end{bmatrix}$.

Then $S^{-1}PS = Q$ and the two permutation matrices P and Q are similar although $P \neq Q$.

9. **False.** See the answer to Question 3.

10. **True.** Let $\lambda^n - p(\lambda)$ be the characteristic polynomial of A. Then by Theorem 2.8, $A^n - p(A) = 0$ or $A^n = p(A)$.

Exercises

1. $\det(\lambda I_n - A) = (\lambda - 1)^2(\lambda - 4) + 4 = \lambda(\lambda - 3)^2$. Thus the characteristic roots of A are 0, 3, 3. A characteristic vector corresponding to 0 is

obtained by solving the system

$$(0I_n - A)x = 0$$

or

$$Ax = 0$$

which yields $x_1 = \frac{1}{3}(2,2,1)$. Now use the method and notation of the proof of Theorem 2.6. The unit vector x_1 together with $x_2 = \dfrac{1}{\sqrt{2}}(1,-1,0)$ and $x_3 = \dfrac{1}{\sqrt{18}}(1,1,-4)$ form an o.n. basis for $V_3(R)$. Let

$$S = \begin{bmatrix} 2/3 & 1/\sqrt{2} & 1/\sqrt{18} \\ 2/3 & -1/\sqrt{2} & 1/\sqrt{18} \\ 1/3 & 0 & -4/\sqrt{18} \end{bmatrix}.$$

Then S is orthogonal and

$$S^T A S = \frac{3}{2} \begin{bmatrix} 0 & 0 & 0 \\ 0 & 1 & 1 \\ 0 & -1 & 3 \end{bmatrix}$$

Let $A_1 = \dfrac{3}{2} \begin{bmatrix} 1 & 1 \\ -1 & 3 \end{bmatrix}$. Now, the characteristic roots of A_1 are 3, 3 and a solution of the system

$$(3I_2 - A_1) = \frac{3}{2} \begin{bmatrix} 1 & -1 \\ 1 & -1 \end{bmatrix} v = 0$$

is $v_1 = \dfrac{1}{\sqrt{2}}(1,1)$, yielding

$$L_1 = \frac{1}{\sqrt{2}} \begin{bmatrix} 1 & 1 \\ 1 & -1 \end{bmatrix}$$

and

$$L_1^T A_1 L_1 = \begin{bmatrix} 3 & -3 \\ 0 & 3 \end{bmatrix}.$$

Thus

$$P^T A P = \begin{bmatrix} 0 & 0 & 0 \\ 0 & 3 & -3 \\ 0 & 0 & 3 \end{bmatrix}$$

where

$$P = S([1] \dot{+} L_1) = \frac{1}{3} \begin{bmatrix} 2 & 2 & 1 \\ 2 & -1 & -2 \\ 1 & -2 & 2 \end{bmatrix}.$$

2. In the solution of Exercise 1 we saw that the rank of $0I_3 - A$ is 2 and therefore, by Theorem 3.3 (a), Chapter 2, any characteristic vector corresponding to 0 is a nonzero multiple of $(2,2,1)$. The other two characteristic roots of A are both equal to 3. The corresponding characteristic vectors are the solutions of the system

$$(3I_3 - A)x = 0.$$

But the rank of the matrix

$$3I_3 - A = \begin{bmatrix} 2 & 0 & 2 \\ 1 & 2 & 0 \\ 0 & 2 & -1 \end{bmatrix}$$

is 2 and therefore all characteristic vectors of A corresponding to 3 are nonzero multiples of $(2, -1, -2)$. Thus we have at most two linearly independent characteristic vectors of A and, by Theorem 2.1, the matrix is not similar to a diagonal matrix.

3. By Theorem 2.4, the set of upper triangular n-square matrices is closed under addition and multiplication. The additive and multiplicative identities, 0 and I_n, are both triangular. The associative and distributive laws hold for all n-square matrices. The addition is clearly commutative and the additive inverse of a triangular matrix A is the matrix $-A$ which also is triangular.

4. By Theorem 2.6, there exists a matrix S such that $S^{-1}AS = B = (b_{ij})$, where B is upper triangular. Since all characteristic roots of A, and therefore of B (see Theorem 1.3), are 0, the diagonal entries of B are 0. Thus $b_{ij} = 0$ whenever $i \leq j$. Denote the (i,j) entry of B^k by $b_{ij}^{(k)}$. We prove by induction on k that $b_{ij}^{(k)} = 0$ for $j = 1, \ldots, i + k - 1, i = 1, \ldots, n$. The statement holds for $k = 1$; assume that $b_{ij}^{(k-1)} = 0$ for $j = 1, \ldots, i + k - 2, i = 1, \ldots, n$. Then

$$b_{ij}^{(k)} = \sum_{t=1}^{n} b_{it}^{(k-1)} b_{tj}.$$

Now, for $t = 1, \ldots, i + k - 2$, $b_{it}^{(k-1)} = 0$, by the induction hypothesis. If $j \leq i + k - 1$ then, for $t = i + k - 1, \ldots, n$, we have $t \geq j$

and therefore $b_{tj} = 0$. Thus for $j \leq i + k - 1$ every summand on the right-hand side is zero and $b_{ij}^{(k)} = 0$. This completes our proof by induction. We observe that $b_{ij}^{(k)} = 0$ for all $j \leq i + n - 1$, $i = 1, \ldots, n$, and therefore $b_{ij}^{(n)} = 0$ for all i, j. Hence $B^n = 0$ and $A^n = (SBS^{-1})^n = SB^nS^{-1} = 0$.

5.   Let A be an indempotent, n-square matrix and let $S^{-1}AS$ be a triangular matrix (see Theorem 2.6) with main diagonal entries $\lambda_1, \ldots, \lambda_n$. These are the characteristic roots of $S^{-1}AS$ and thus, by Theorem 1.3, of A. Now, $(S^{-1}AS)^2 = S^{-1}A^2S$ is triangular with main diagonal entries λ_1^2, $\ldots, \lambda_n^2$. But $S^{-1}A^2S = S^{-1}AS$ and therefore $\lambda_j = \lambda_j^2$, $j = 1, \ldots, n$, i.e., $\lambda_j = 0$ or 1, $j = 1, \ldots, n$.

6.   Let $G = \{g_1, \ldots, g_n\}$ be any o.n. basis of V, and let $U = (u_{ij})$ be a unitary matrix such that

$$U^*[T]_G{}^G U = B$$

is triangular (see Theorem 2.6). Define

$$h_j = \sum_{i=1}^{n} \bar{u}_{ji}g_i, \quad j = 1, \ldots, n.$$

Then

$$(h_s,h_t) = \left(\sum_{i=1}^{n} \bar{u}_{si}g_i, \sum_{k=1}^{n} \bar{u}_{tk}g_k \right)$$

$$= \sum_{i,k=1}^{n} \bar{u}_{si}u_{tk}(g_i,g_k)$$

$$= \sum_{i=1}^{n} \bar{u}_{si}u_{ti}$$

$$= (UU^*)_{ts}$$

$$= \delta_{st}.$$

Thus $\{h_1, \ldots, h_n\}$ is an o.n. basis of V. Also, by Def. 4.1, Chap. 1, $[I_V]_H{}^G = U^*$ and thus $[I_V]_G{}^H = U$ and

$$B = U^*[T]_G{}^G U = [I_V]_H{}^G[T]_G{}^G[I_V]_G{}^H = [T]_H{}^H$$

is triangular.

Section 3.3

Quiz

1. **True.**   The conjugate transpose of the matrix is $\begin{bmatrix} 1-i & -i \\ -i & 1+i \end{bmatrix}$ and

$$\begin{bmatrix} 1+i & i \\ i & 1-i \end{bmatrix}\begin{bmatrix} 1-i & i \\ -i & 1+i \end{bmatrix} = \begin{bmatrix} 3 & 0 \\ 0 & 3 \end{bmatrix}$$
$$= \begin{bmatrix} 1-i & -i \\ -i & 1+i \end{bmatrix}\begin{bmatrix} 1+i & i \\ i & 1-i \end{bmatrix}.$$

2. **True.**   If A and B are normal and have the same characteristic vectors, then they are simultaneously diagonalizable (see Theorems 3.1 and 2.1). Thus there exists a matrix U such that

$$U^{-1}AU = \text{diag}\,(c_1, \ldots, c_n) \quad \text{and} \quad U^{-1}BU = \text{diag}\,(d_1, \ldots, d_n)$$

and

$$\begin{aligned} AB &= U\,\text{diag}\,(c_1, \ldots, c_n)U^{-1}U\,\text{diag}\,(d_1, \ldots, d_n)U^{-1} \\ &= U\,\text{diag}\,(c_1 d_1, \ldots, c_n d_n)U^{-1} \\ &= U\,\text{diag}\,(d_1, \ldots, d_n)U^{-1}U\,\text{diag}\,(c_1, \ldots, c_n)U^{-1} \\ &= BA. \end{aligned}$$

3. **False.**   The matrix $\dfrac{1}{5}\begin{bmatrix} 3 & 4 \\ 4 & -3 \end{bmatrix}$ is orthogonal.

4. **True.**   Let A be normal and $A^k = 0$. Let $U^*AU = D$, where D is a diagonal matrix. Then

$$D^k = (U^*AU)^k = U^*A^kU = 0$$

and thus $D = 0$. Therefore

$$A = UDU^* = 0.$$

Conversely, if $A = 0$, then A is normal and nilpotent.

5. **True.**   If A is both hermitian and skew-hermitian then, by Theorem 3.5, its characteristic roots are both real and imaginary and thus they are all equal to 0. Thus A is similar to a zero matrix and therefore $A = 0$.

6. **False.**   For example,

$$A = \begin{bmatrix} 2 & 1 \\ 1 & 0 \end{bmatrix} \quad \text{and} \quad B = \begin{bmatrix} 0 & -1 \\ 1 & 0 \end{bmatrix}$$

are both normal (*A* being hermitian and *B* skew-hermitian) but

$$(AB)(AB)^* = \begin{bmatrix} 1 & -2 \\ 0 & -1 \end{bmatrix} \begin{bmatrix} 1 & 0 \\ -2 & -1 \end{bmatrix} = \begin{bmatrix} 5 & 2 \\ 2 & 1 \end{bmatrix}$$

$$\neq (AB)^*(AB) = \begin{bmatrix} 1 & -2 \\ -2 & 5 \end{bmatrix}.$$

7. **True.** For, if *U* is a unitary matrix such that $U^*AU = \text{diag}(\lambda_1, \ldots, \lambda_n)$, then

$$\begin{aligned} U^*(I_n - A)U &= U^*I_nU - U^*AU \\ &= I_n - \text{diag}(\lambda_1, \ldots, \lambda_n) \\ &= \text{diag}(1 - \lambda_1, \ldots, 1 - \lambda_n) \end{aligned}$$

and $I_n - A$ is normal. Conversely, if $I_n - A$ is normal then, by the first part, $I_n - (I_n - A) = A$ is normal.

8. **False.** If *A* is skew-hermitian then, by Theorem 3.5, its characteristic roots $\lambda_1, \ldots, \lambda_n$ are pure imaginary. But then the characteristic roots of $I_n - A$ are $1 - \lambda_1, \ldots, 1 - \lambda_n$ (see Exercise 5, Sect. 3.1) which cannot be pure imaginary and therefore, by Theorem 3.5, $I_n - A$ cannot be skew-hermitian. The converse is also false since $A = I_n - (I_n - A)$.

9. **False.** Let $B = \begin{bmatrix} 1 & 1 \\ 0 & 1 \end{bmatrix}$ and $A = I_n$. Then *A* is normal while *B* is not unitary. Nevertheless

$$B^{-1}AB = B^{-1}I_nB = I_n$$

which is diagonal.

10. **False.** Any *n*-square matrix *A* is equal to $H + iK$ where $H = (A + A^*)/2$ and $K = (A - A^*)/2i$ are both hermitian.

Exercises

1. Let $A = \begin{bmatrix} a_{11} & a_{12} \\ a_{21} & a_{22} \end{bmatrix}$ be any real orthogonal matrix. Then $a_{11}^2 + a_{12}^2 = a_{11}^2 + a_{21}^2 = a_{21}^2 + a_{22}^2 = a_{12}^2 + a_{22}^2 = 1$. It follows that

$$a_{11}^2 = a_{22}^2 = 1 - a_{12}^2 = 1 - a_{21}^2.$$

Set $a_{11} = \cos\theta$ and $a_{12} = \sin\theta$ (i.e., $\theta = \cos^{-1}a_{11}$ if $a_{12} \geq 0$ and $\theta = 2\pi - \cos^{-1}a_{11}$ if $a_{12} < 0$). If $a_{11} = a_{22}$, then $a_{11}a_{21} + a_{12}a_{22} = 0$ implies that $a_{12} = -a_{12}$ and we obtain

$$A = \begin{bmatrix} \cos\theta & \sin\theta \\ -\sin\theta & \cos\theta \end{bmatrix}.$$

If $a_{11} = -a_{22}$ then we must have $a_{12} = a_{21}$, and

$$A = \begin{bmatrix} \cos\theta & \sin\theta \\ \sin\theta & -\cos\theta \end{bmatrix}.$$

2. Let $\lambda_1, \ldots, \lambda_n$ be the characteristic roots of U. Then

$$\begin{aligned} |\operatorname{tr} U| = |\lambda_1 + \cdots + \lambda_n| \\ \leq |\lambda_1| + \cdots + |\lambda_n| \\ = n, \end{aligned}$$

by Theorem 3.5. If $|\operatorname{tr} U| = n$, then $\lambda_1 = \cdots = \lambda_n = c$ for some c, $|c| = 1$. But then all the roots of the normal matrix $\frac{1}{c} U$ are equal to 1 and $\frac{1}{c} U = I_n$. Conversely, if $U = cI_n$, then $\operatorname{tr} U = nc$ and $|\operatorname{tr} U| = n$.

3. If A is normal, then U^*AU is diagonal for some unitary matrix U and $U^*A^kU = (U^*AU)^k$ is also diagonal. Thus A^k is unitarily diagonizable and therefore, by Theorem 3.1, it is normal. Alternatively, we note that $(A^k)^* = (A^*)^k$ and that $AA^* = A^*A$ implies $A^k(A^k)^* = A^k(A^*)^k = (A^*)^kA^k = (A^k)^*A^k$.

4. Let $H = \begin{bmatrix} a & c \\ c & b \end{bmatrix}.$ Then

$$\det (H) = ab - c^2 \leq ab + c^2 = \operatorname{per} (H)$$

Also

$$\tfrac{1}{2}(\lambda_1{}^2 + \lambda_2{}^2) = \tfrac{1}{2}(\lambda_1 + \lambda_2)^2 - \lambda_1\lambda_2 = \tfrac{1}{2}(a + b)^2 - (ab - c^2),$$

since $\lambda_1 + \lambda_2 = \operatorname{tr} (H) = a + b$ and $\lambda_1\lambda_2 = \det (H)$. Therefore

$$\tfrac{1}{2}(\lambda_1{}^2 + \lambda_2{}^2) = \tfrac{1}{2}(a^2 + b^2) + c^2.$$

But

$$\tfrac{1}{2}(a^2 + b^2) \geq ab \text{ (arithmetic-geometric mean inequality)}$$

and thus

$$\tfrac{1}{2}(\lambda_1{}^2 + \lambda_2{}^2) \geq ab + c^2 = \operatorname{per} (H).$$

5. If $\det (H) = \operatorname{per} (H)$, then $ab - c^2 = ab + c^2$, $c = 0$ and $H = \operatorname{diag} (a,b)$. If $\operatorname{per} (H) = \tfrac{1}{2}(\lambda_1{}^2 + \lambda_2{}^2)$, then $\tfrac{1}{2}(a^2 + b^2) = ab$ (see solution of the

preceding question), i.e., $\frac{1}{2}(a - b)^2 = 0$. Thus we must have $a = b$ and

$$H = \begin{bmatrix} a & c \\ c & a \end{bmatrix}.$$

6. If A and B have two linearly independent characteristic vectors in common, then, by Theorem 2.1,

$$S^{-1}AS = \begin{bmatrix} \lambda_1 & 0 \\ 0 & \lambda_2 \end{bmatrix} \quad \text{and} \quad S^{-1}BS = \begin{bmatrix} \mu_1 & 0 \\ 0 & \mu_2 \end{bmatrix}$$

and

$$\begin{aligned} AB &= S \begin{bmatrix} \lambda_1 & 0 \\ 0 & \lambda_2 \end{bmatrix} S^{-1} \; S \begin{bmatrix} \mu_1 & 0 \\ 0 & \mu_2 \end{bmatrix} S^{-1} \\ &= S \begin{bmatrix} \lambda_1\mu_1 & 0 \\ 0 & \lambda_2\mu_2 \end{bmatrix} S^{-1} \\ &= S \begin{bmatrix} \mu_1 & 0 \\ 0 & \mu_2 \end{bmatrix} S^{-1} \; S \begin{bmatrix} \lambda_1 & 0 \\ 0 & \lambda_2 \end{bmatrix} S^{-1} \\ &= BA. \end{aligned}$$

If neither A nor B have two linearly independent characteristic vectors then, by Theorems 2.2 and 2.6, we can find a matrix U such that

$$U^{-1}AU = \begin{bmatrix} \lambda_1 & a \\ 0 & \lambda_1 \end{bmatrix} \quad \text{and} \quad U^{-1}BU = \begin{bmatrix} \mu_1 & b \\ 0 & \mu_1 \end{bmatrix}.$$

But then

$$\begin{aligned} AB &= U \begin{bmatrix} \lambda_1 & a \\ 0 & \lambda_1 \end{bmatrix} U^{-1} \; U \begin{bmatrix} \mu_1 & b \\ 0 & \mu_1 \end{bmatrix} U^{-1} \\ &= U \begin{bmatrix} \lambda_1\mu_1 & \lambda_1 b + a\mu_1 \\ 0 & \lambda_1\mu_1 \end{bmatrix} U^{-1} \\ &= U \begin{bmatrix} \mu_1 & b \\ 0 & \mu_1 \end{bmatrix} U^{-1} \; U \begin{bmatrix} \lambda_1 & a \\ 0 & \lambda_1 \end{bmatrix} U^{-1} \\ &= BA. \end{aligned}$$

Note that this property does not hold for n-square matrices, $n > 2$ (see solution of Exercise 10).

7. Since A is normal, there exists a unitary matrix U such that $U^*AU = \text{diag}(\lambda_1, \ldots, \lambda_n)$. Let $B = U \, \text{diag}(\lambda_1^{1/2}, \ldots, \lambda_n^{1/2}) U^*$. Then

$$\begin{aligned} B^2 &= U \, \text{diag}(\lambda_1^{1/2}, \ldots, \lambda_n^{1/2}) U^* U (\lambda_1^{1/2}, \ldots, \lambda_n^{1/2}) U^* \\ &= U \, \text{diag}(\lambda_1, \ldots, \lambda_n) U^* \\ &= A. \end{aligned}$$

8. By the induction hypothesis, there exists a unitary $(n-1)$-square matrix S_1 such that $S_1 {}^* A_1 S_1$ and $S_1 {}^* B_1 S_1$ are triangular. Let $U = S(I_1 + S_1)$. Then U is unitary and

$$U^* A U = \left[\begin{array}{c|c} \lambda_1 & zS_1 \\ \hline 0 & S_1{}^* A_1 S_1 \end{array}\right], \quad U^* B U = \left[\begin{array}{c|c} \mu_1 & wS_1 \\ \hline 0 & S_1{}^* B_1 S_1 \end{array}\right]$$

are both triangular.

9. Every square matrix commutes with itself (or with I_n) and therefore can be unitarily triangularized.

10. The matrices $A = \begin{bmatrix} 0 & 1 & 0 \\ 0 & 0 & 0 \\ 0 & 0 & 0 \end{bmatrix}$ and $B = \begin{bmatrix} 0 & 0 & 0 \\ 0 & 0 & 1 \\ 0 & 0 & 0 \end{bmatrix}$

are simultaneously triangularizable, e.g., by I_3. However,

$$AB = \begin{bmatrix} 0 & 0 & 1 \\ 0 & 0 & 0 \\ 0 & 0 & 0 \end{bmatrix} \neq BA = 0_{3,3}.$$

Section 3.4

Quiz

1. **False.** For example, if $x = (1, 1)$ and $A = \begin{bmatrix} 1 & -3 \\ 0 & 1 \end{bmatrix}$, then $(Ax,x) = -1$. Of course, if A is a symmetric matrix with positive characteristic roots then $(Ax,x) > 0$ for any nonzero x (see Theorem 4.12).

2. **False.** Let $A = \begin{bmatrix} 0 & 1 \\ 1 & 0 \end{bmatrix}$ and $B = \begin{bmatrix} 0 & 2 \\ 1 & 0 \end{bmatrix}$. Then A, $AB = \text{diag}(1,2)$, and $BA = \text{diag}(2,1)$ are all hermitian although B is not a hermitian matrix.

3. **True.** As a matter of fact, this statement is true for any diagonalizable matrix. For, by Theorem 3.1, S is similar to a diagonal matrix D. Now if all the characteristic roots of S are equal to α then $D = \alpha I_n$ and $S = U^{-1} \alpha I_n U = \alpha I_n$. Since S is skew-hermitian α is pure imaginary [see Theorem 4.1 (c)].

4. **True.** This follows directly from Theorem 4.6 or from Exercise 2 below and Theorem 3.2 (b).

5. **False.** The matrix has negative principal subdeterminants of order 2 and thus, by Theorem 4.11, it cannot be positive semidefinite.

6. **True.** For, if $H = U^* \text{diag}(\lambda_1, \ldots, \lambda_n) U$, where U is unitary, and $\lambda_j > 0$, $j = 1, \ldots, n$, then there are exactly 2^n distinct hermitian matrices of the form $A = U^* \text{diag}(\pm \sqrt{\lambda_1}, \ldots, \pm \sqrt{\lambda_n}) U$ corresponding to the 2^n distinct choices of $+$ and $-$ in diag $(\pm \sqrt{\lambda_1}, \ldots, \pm \sqrt{\lambda_n})$. For each of these matrices $A^2 = U^*(\lambda_1, \ldots, \lambda_n) U = H$.

7. **True.** Let $A = UH$, where U is unitary and H positive semidefinite hermitian. If $A = 0$, i.e., $UH = 0$, then $H = 0$ (since U is nonsingular) and all the characteristic roots of H, i.e., the singular values of A, are 0. Conversely, if all the singular values of A are 0, then, by Theorem 3.4, $H = 0$ and therefore $A = UH = 0$.

8. **False.** For example, if $H = I_n$ then, for any unitary n-square matrix U, we have $U^*I_nU = \text{diag}(1, \ldots, 1)$.

9. **False.** If it were true then H, by Theorem 4.12, would be positive definite, which is not the case because $\det(H) = 0$ (see Theorem 4.11). In fact, if $x = (0, 1, -1)$, then $(Hx,x) = 0$ although $x \neq 0$.

10. **False.** For, if $B = UA$ where U is any unitary matrix, then $A^*A = B^*B$.

Exercises

1. If B is a real matrix with real characteristic roots then, by Theorem 2.6, there exists a real orthogonal matrix P such that P^TBP is triangular. If B happens to be symmetric, then P^TBP is symmetric as well and therefore diagonal. Thus a real symmetric n-square matrix has n orthonormal real characteristic vectors (see Theorem 2.1). Now, Theorem 4.14 can be proved by following step-by-step the proof of Theorem 4.13. The orthonormal characteristic vectors $x_1, \ldots, x_n$, of A^TA, can be chosen to be real and then $z_1, \ldots, z_n$ are real. Thus U, V are real orthogonal and H,K are real symmetric. The proof of the second part of the theorem is identical with the proof of Theorem 4.13.

2. By Theorem 4.6, if A is a real skew-symmetric n-square matrix, then it is similar to a real skew-symmetric matrix of the form

$$\begin{bmatrix} 0 & a_1 \\ -a_1 & 0 \end{bmatrix} + \cdots + \begin{bmatrix} 0 & a_m \\ -a_m & 0 \end{bmatrix} + 0_{n-2m,n-2m},$$

where $a_i \neq 0$, $i = 1, \ldots, m$. Thus A has $n - 2m$ zero characteristic roots and, by Theorem 3.4, the rank of A is $n - (n - 2m) = 2m$, an even number.

3. Let $A = UH$ where U is unitary and H positive semidefinite hermitian. Then the singular values of A are the characteristic roots of H. But $H^2 = A^*A = V^*K^2V$ and thus H^2 and K^2 are similar. Hence the (nonnegative) square roots of the characteristic roots of H^2, i.e., the singular values of A, are the characteristic roots of K.

4. The rank of the symmetric matrix J is 1 and therefore, by Theorem 3.4, $n - 1$ roots of J are equal to 0. Now tr $(J) = n$ and thus the single nonzero characteristic root of J is equal to n. By Exercises 9 and 4, Sect. 3.1, $n - 1$ of the characteristic roots of $I_n + xJ$ are equal to 1 and one is equal to $1 + nx$. Thus $I_n + xJ$ is positive definite if and only if $1 + nx > 0$, i.e., $x > -1/n$.

5. We compute the characteristic polynomial of H,

$$\lambda^3 - 18\lambda^2 + 81\lambda.$$

Hence the characteristic roots of H are 9, 9 and 0. The corresponding characteristic vectors are obtained by solving

$$(9I_3 - H)x = \begin{bmatrix} 1 & 2 & 2 \\ 2 & 4 & 4 \\ 2 & 4 & 4 \end{bmatrix} \begin{bmatrix} x_1 \\ x_2 \\ x_3 \end{bmatrix} = 0,$$

and $Hx = 0$. We find that an orthonormal set of characteristic vectors is $\{\frac{1}{3}(2,1,-2), \frac{1}{3}(2,-2,1), \frac{1}{3}(1,2,2)\}$.

Let $U = \dfrac{1}{3} \begin{bmatrix} 2 & 2 & 1 \\ 1 & -2 & 2 \\ 2 & 1 & 2 \end{bmatrix}$. Then $H = U$ diag $(9,9,0)U^*$ and

$$H^{\frac{1}{2}} = U \text{ diag } (3,3,0) \ U^* = \frac{1}{3} \begin{bmatrix} 8 & -2 & -2 \\ -2 & 5 & -4 \\ -2 & -4 & 5 \end{bmatrix} = \frac{1}{3} H$$

6. The characteristic polynomial of A^*A is $\lambda^3 - 6\lambda^2 + 8\lambda$. Thus the characteristic roots of A^*A are 2, 4 and 0. We compute the characteristic vectors of unit length corresponding to these roots:

$$x_1 = (0,1,0), \quad x_2 = \frac{1}{\sqrt{2}}(1,0,1), \quad x_3 = \frac{1}{\sqrt{2}}(1,0,-1).$$

Let

$$X = \frac{1}{\sqrt{2}} \begin{bmatrix} 0 & 1 & 1 \\ \sqrt{2} & 0 & 0 \\ 0 & 1 & -1 \end{bmatrix}.$$

Let $z_1 = Ax_1/\sqrt{2} = \dfrac{1}{\sqrt{2}}(1,-1,0)$, $z_2 = Ax_2/2 = \dfrac{1}{\sqrt{2}}(1,1,0)$ and set

$$Z = \frac{1}{\sqrt{2}} \begin{bmatrix} 1 & 1 & 0 \\ -1 & 1 & 0 \\ 0 & 0 & \sqrt{2} \end{bmatrix}.$$

Then $A = UH$, where

$$U = ZX^* = \frac{1}{2}\begin{bmatrix} 1 & \sqrt{2} & 1 \\ 1 & -\sqrt{2} & 1 \\ \sqrt{2} & 0 & -\sqrt{2} \end{bmatrix}$$

and

$$H = X \text{ diag } (\sqrt{2},2,0)X^* = \begin{bmatrix} 1 & 0 & 1 \\ 0 & \sqrt{2} & 0 \\ 1 & 0 & 1 \end{bmatrix}.$$

By Example 3, the characteristic roots of AA^* are 2, 4 and 0. Again, we find an orthonormal set of characteristic vectors

$$x_1' = \frac{1}{\sqrt{2}}(1,-1,0), \quad x_2' = \frac{1}{\sqrt{2}}(1,1,0), \quad x_3' = (0,0,1)$$

and we compute

$$z_1' = A^*x_1'/\sqrt{2} = (0,1,0), \quad z_2' = A^*x_2'/2 = \frac{1}{\sqrt{2}}(1,0,1).$$

We then set

$$X' = \frac{1}{\sqrt{2}}\begin{bmatrix} 1 & 1 & 0 \\ -1 & 1 & 0 \\ 0 & 0 & \sqrt{2} \end{bmatrix} \quad \text{and} \quad Z' = \frac{1}{\sqrt{2}}\begin{bmatrix} 0 & 1 & 1 \\ \sqrt{2} & 0 & 0 \\ 0 & 1 & -1 \end{bmatrix}.$$

Then $A^* = V^*K$ or $A = KV$, where

$$K = X' \text{ diag } (\sqrt{2},2,0)X'^* = \frac{1}{2}\begin{bmatrix} 2+\sqrt{2} & 2-\sqrt{2} & 0 \\ 2-\sqrt{2} & 2+\sqrt{2} & 0 \\ 0 & 0 & 0 \end{bmatrix}$$

and

$$V = X'Z'^* = \frac{1}{2}\begin{bmatrix} 1 & \sqrt{2} & 1 \\ 1 & -\sqrt{2} & 1 \\ \sqrt{2} & 0 & -\sqrt{2} \end{bmatrix}.$$

Section 3.5

Quiz

1. **False.**   Let $A = \text{diag} (1,-1,-1)$. Then $\det (A) = 1$ and

$$\left[\frac{\text{tr} (A)}{n} \right]^n = \left(-\frac{1}{3} \right)^3 = -\frac{1}{27}.$$

2. **True.** $\det (A[1|1]) = 10 > 0$, $\det (A[1,2|1,2]) = 46 > 0$, $\det (A) = 215$. Apply Theorem 5.10.

3. **True.**   By the second part of Theorem 5.5, the characteristic roots of the matrix lie in the closed region consisting of all the disks $|z - 3| \leq 2$, $|z + 4| \leq 3$, $|z - 3| \leq 2$ and $|z - 5| \leq 4$. None of these disks contains 0 and therefore all the characteristic roots are nonzero and the matrix is nonsingular [see Theorem 1.5 (a) and Theorem 6.4, Chap. 2].

4. **False.**   For example, the matrix $A = \begin{bmatrix} 0 & 1 \\ 1 & 0 \end{bmatrix}$ is nonsingular although here

$$|a_{ii}| < \sum_{\substack{t = 1 \\ t \neq i}}^{2} |a_{it}|, \qquad i = 1,2.$$

5. **False.**   The characteristic roots of the matrix $\begin{bmatrix} 5 & -6 \\ 1 & 0 \end{bmatrix}$ are 2 and 3. They both lie outside the Geršgorin disk $|z - 0| \leq 1$.

6. **True.**   In fact, equality holds.

7. **True.**   Let λ_1 be the largest characteristic root of H. Then, by Theorem 5.2 (a),

$$\lambda_1 \geq (Hx,x)$$

for any unit vector x. In particular, if e_1 is the standard vector with 1 in its first coordinate, then

$$\lambda_1 \geq (He_1,e_1) = h_{11} = 0$$

and since H is nonsingular $\lambda_1 \neq 0$ and thus λ_1 must be positive.

8. **True.**   All the three Geršgorin disks $|z + 3| \leq 2$, $|z + 4| \leq 3$ and $|z + 5| \leq 4$ lie to the left of the imaginary axis. Hence all the characteristic roots of the matrix lie to the left of the imaginary axis, and thus have negative real parts.

9. **True.**   This follows directly from Theorem 5.4: here $|a_{ii}| = n$ and $P_i = n - 1$, $i = 1, \ldots , n$.

10. **False.** Let $H = \begin{bmatrix} 5 & 0 \\ 0 & 1 \end{bmatrix}$ and $K = \begin{bmatrix} 4 & 0 \\ 0 & 3 \end{bmatrix}$. Then $\lambda_1 = 5 > \mu_1 = 4$. However, if $x = (1,1)$, then $(Hx,x) = 6 < (Kx,x) = 7$.

Exercises

1. Let $e_1, \ldots, e_n$ be the standard basis of $V_n(R)$. Then

$$(He_i,e_i) = h_{ii}, \quad i = 1, \ldots, n,$$

and therefore, by Theorem 5.2,

$$\lambda_n \le h_{ii} \le \lambda_1, \quad i = 1, \ldots, n.$$

2. By Theorem 5.5, we have

$$|\lambda_t - a_{ii}| \le P_i$$

for some i, $1 \le i \le n$. But

$$|\lambda_t - a_{ii}| \ge |a_{ii}| - |\lambda_t|$$

and therefore

$$|a_{ii}| - |\lambda_t| \le P_i$$

or

$$|\lambda_t| \ge |a_{ii}| - P_i$$

for some i. Thus

$$|\lambda_t| \ge \min_i \left(|a_{ii}| - P_i \right).$$

3. If λ_j is a characteristic root of A, then, by Theorem 5.5,

$$|\lambda_j - a_{ii}| \le P_i = \sum_{t=1}^{n} |a_{it}| - |a_{ii}|$$

for some i. Now,

$$|\lambda_j| - |a_{ii}| \le |\lambda_j - a_{ii}|.$$

Therefore

$$|\lambda_j| - |a_{ii}| \le \sum_{t=1}^{n} |a_{it}| - |a_{ii}|$$

and

$$|\lambda_j| \le \sum_{t=1}^{n} |a_{it}|$$

for some i. Hence

$$|\lambda_j| \le \max_i \sum_{t=1}^{n} |a_{it}|.$$

Now, applying the above bound,

$$|\det (A)| = \prod_{j=1}^{n} |\lambda_j| \le \left(\max_i \sum_{t=1}^{n} |a_{it}|\right)^n.$$

4. We compute

$$B = \tfrac{1}{2}(A + A^*) = \begin{bmatrix} 3 & 3 & 3 \\ 3 & 3 & 3 \\ 3 & 3 & 3 \end{bmatrix}.$$

The matrix B is a symmetric matrix of rank 1. Now, tr $(B) = 9$ and therefore its characteristic roots are 9, 0 and 0. Thus, by Theorem 5.3,

$$0 \le \text{Re}\ (\lambda_t) \le 9$$

for any characteristic root λ_t of A. (As a matter of fact, the characteristic roots of A are 8, $1 \pm i\sqrt{23}/2$.)

5. The matrix AA^* is hermitian positive semidefinite. If we apply Theorem 5.7 to AA^* we have

$$|\det (A)|^2 = \det (A) \det (A^*) = \det (AA^*)$$
$$\le \prod_{i=1}^{n} (AA^*)_{ii} = \prod_{i=1}^{n} \sum_{j=1}^{n} a_{ij}\bar{a}_{ij}$$
$$= \prod_{i=1}^{n} \sum_{j=1}^{n} |a_{ij}|^2.$$

6. Let $\lambda_1 \geq \lambda_2 \geq \lambda_3$ be the three characteristic roots of A. We are going to use Theorem 5.9 applied first to the submatrix $[a_{11}]$ and then to the submatrix $A[1,2|1,2]$. We know that $\lambda_1 \geq a_{11} > 0$ so that $\lambda_1 > 0$. If $\mu_1 \geq \mu_2$ are the characteristic roots of $A[1,2|1,2]$ we know that $\lambda_1 \geq \mu_1 \geq \lambda_2 \geq \mu_2 \geq \lambda_3$. Then $\mu_1 > 0$ because $a_{11} > 0$. Also, by hypothesis, $\mu_1\mu_2 = \det(A[1,2|1,2]) < 0$. Hence $\mu_2 < 0$ so that $\lambda_3 < 0$. But $\det(A) > 0$ so $\lambda_1\lambda_2\lambda_3 > 0$. It follows, now that we know $\lambda_1 > 0$ and $\lambda_3 < 0$, that $\lambda_2 < 0$.
7. By Theorem 2.6, there exists a unitary matrix U such that $G = (g_{ij}) = U^*AU$ is upper triangular. If $\lambda_1, \ldots, \lambda_n$ denote the characteristic roots of A suitably ordered, then $\lambda_i = g_{ii}, i = 1, \ldots, n$. Now,

$$U^*(A + A^*)(A + A^*)^*U = (G + G^*)(G + G^*)^*$$

and, by Theorem 1.5 (c),

$$\text{tr}\,[(A + A^*)(A + A^*)^*] = \text{tr}\,[(G + G^*)(G + G^*)^*].$$

Hence, by Theorem 1.6 (d),

$$
\begin{aligned}
\sum_{i,j=1}^{n} |a_{ij} + \bar{a}_{ji}|^2 &= \sum_{i,j=1}^{n} |g_{ij} + \bar{g}_{ji}|^2 \\
&= \sum_{i=1}^{n} |\lambda_i + \bar{\lambda}_i|^2 + \sum_{i<j} |g_{ij} + \bar{g}_{ji}|^2 \\
&\geq \sum_{i=1}^{n} |\lambda_i + \bar{\lambda}_i|^2 \\
&= \sum_{i=1}^{n} |2\,\text{Re}\,(\lambda_i)|^2.
\end{aligned}
$$

Hence,

$$n^2 \max_{i,j} |a_{ij} + \bar{a}_{ji}|^2 \geq 4 \sum_{i=1}^{n} |\text{Re}\,(\lambda_i)|^2.$$

8. Suppose that a symmetric nonzero matrix has no positive roots. Now, a nonzero symmetric matrix must have at least one nonzero root and therefore the trace of the matrix must be negative. But this is impossible if all the entries of the matrix are nonnegative.
9. The centers of the Geršgorin disks corresponding to the rows of the matrix are $1 - 3i$, $2 + 4i$ and $1 + 3i$; the radii are $1 + \sqrt{2}$, $2\sqrt{2}$ and

$1 + \sqrt{2}$ respectively. Now, $1 + \sqrt{2} < 3$ and $2\sqrt{2} < 4$, and therefore none of the disks intersects the real axis and, by Theorem 5.5, the matrix has no real roots.

10. Define T_A and T_B by $T_A x = Ax$ and $T_B x = Bx$ as hermitian transformations on $V_n(R)$. Then obviously $T_{A+B} = T_A + T_B$. Let $u_1, \ldots, u_n$ be orthonormal characteristic vectors of T_A corresponding respectively to $\lambda_1, \ldots, \lambda_n$. Let $W = \langle u_k, \ldots, u_n \rangle$ so that $\dim W = n - k + 1$. Then by Theorem 5.8,

$$
\begin{aligned}
\sigma_k &\le \max_{x \in W, \|x\| = 1} (T_{A+B} x, x) \\
&= \max_{x \in W, \|x\| = 1} \{ (T_A x, x) + (T_B x, x) \} \\
&\le \max_{x \in W, \|x\| = 1} (T_A x, x) \;+\; \max_{x \in W, \|x\| = 1} (T_B x, x) \\
&\le \lambda_k + \mu_1.
\end{aligned}
$$

On the other hand, let $W = \langle u_1, \ldots, u_{n-k+1} \rangle$, $\dim W = n - k + 1$. Then

$$
\begin{aligned}
\sigma_{n-k+1} &\ge \min_{x \in W, \|x\| = 1} (T_{A+B} x, x) \\
&\ge \min_{x \in W, \|x\| = 1} (T_A x, x) \;+\; \min_{x \in W, \|x\| = 1} (T_B x, x) \\
&\ge \lambda_{n-k+1} + \mu_n.
\end{aligned}
$$

11. Let $\mu_1 \ge \cdots \ge \mu_k$ be the characteristic roots of B. Then, by Theorem 5.9,

$$
\lambda_s \ge \mu_s \ge \lambda_{n-k+s}, \quad s = 1, \ldots, k.
$$

Hence

$$
\prod_{j=1}^{k} \lambda_{n-j+1} \le \prod_{j=1}^{k} \mu_j \le \prod_{j=1}^{k} \lambda_j,
$$

so that

$$
\prod_{j=1}^{k} \lambda_{n-j+1} \le \det (B) \le \prod_{j=1}^{k} \lambda_j.
$$

12. Let $x_1, \ldots, x_n$ be a completion of $x_1, \ldots, x_k$ to an orthonormal basis.
Let A be the n-square matrix whose (i,j) entry is (Tx_i,x_j), $i,j = 1,$
$\ldots, n$. Then, according to Theorem 4.5, A is positive definite hermitian
and has the same characteristic roots as T. Also, the matrix B whose
(i,j) entry is (Tx_i,x_j), $i,j = 1, \ldots, k$, is a principal submatrix of A.
We can thus apply the result in the preceding exercise.

Index

A CATALOG OF SELECTED
DOVER BOOKS
IN SCIENCE AND MATHEMATICS

A CATALOG OF SELECTED

DOVER BOOKS

IN SCIENCE AND MATHEMATICS

QUALITATIVE THEORY OF DIFFERENTIAL EQUATIONS, V.V. Nemytskii and V.V. Stepanov. Classic graduate-level text by two prominent Soviet mathematicians covers classical differential equations as well as topological dynamics and erqodic theory. Bibliographies. 523pp. 5⅜ × 8½. 65954-2 Pa. $10.95

MATRICES AND LINEAR ALGEBRA, Hans Schneider and George Phillip Barker. Basic textbook covers theory of matrices and its applications to systems of linear equations and related topics such as determinants, eigenvalues and differential equations. Numerous exercises. 432pp. 5⅜ × 8½. 66014-1 Pa. $8.95

QUANTUM THEORY, David Bohm. This advanced undergraduate-level text presents the quantum theory in terms of qualitative and imaginative concepts, followed by specific applications worked out in mathematical detail. Preface. Index. 655pp. 5⅜ × 8½. 65969-0 Pa. $10.95

ATOMIC PHYSICS (8th edition), Max Born. Nobel laureate's lucid treatment of kinetic theory of gases, elementary particles, nuclear atom, wave-corpuscles, atomic structure and spectral lines, much more. Over 40 appendices, bibliography. 495pp. 5⅜ × 8½. 65984-4 Pa. $11.95

ELECTRONIC STRUCTURE AND THE PROPERTIES OF SOLIDS: The Physics of the Chemical Bond, Walter A. Harrison. Innovative text offers basic understanding of the electronic structure of covalent and ionic solids, simple metals, transition metals and their compounds. Problems. 1980 edition. 582pp. 6⅛ × 9¼. 66021-4 Pa. $14.95

BOUNDARY VALUE PROBLEMS OF HEAT CONDUCTION, M. Necati Özisik. Systematic, comprehensive treatment of modern mathematical methods of solving problems in heat conduction and diffusion. Numerous examples and problems. Selected references. Appendices. 505pp. 5⅜ × 8½. 65990-9 Pa. $11.95

A SHORT HISTORY OF CHEMISTRY (3rd edition), J.R. Partington. Classic exposition explores origins of chemistry, alchemy, early medical chemistry, nature of atmosphere, theory of valency, laws and structure of atomic theory, much more. 428pp. 5⅜ × 8½. (Available in U.S. only) 65977-1 Pa. $10.95

A HISTORY OF ASTRONOMY, A. Pannekoek. Well-balanced, carefully reasoned study covers such topics as Ptolemaic theory, work of Copernicus, Kepler, Newton, Eddington's work on stars, much more. Illustrated. References. 521pp. 5⅜ × 8½. 65994-1 Pa. $11.95

PRINCIPLES OF METEOROLOGICAL ANALYSIS, Walter J. Saucier. Highly respected, abundantly illustrated classic reviews atmospheric variables, hydrostatics, static stability, various analyses (scalar, cross-section, isobaric, isentropic, more). For intermediate meteorology students. 454pp. 6⅛ × 9¼. 65979-8 Pa. $12.95

RELATIVITY, THERMODYNAMICS AND COSMOLOGY, Richard C. Tolman. Landmark study extends thermodynamics to special, general relativity; also applications of relativistic mechanics, thermodynamics to cosmological models. 501pp. 5⅜ × 8½. 65383-8 Pa. $11.95

APPLIED ANALYSIS, Cornelius Lanczos. Classic work on analysis and design of finite processes for approximating solution of analytical problems. Algebraic equations, matrices, harmonic analysis, quadrature methods, much more. 559pp. 5⅜ × 8½. 65656-X Pa. $11.95

SPECIAL RELATIVITY FOR PHYSICISTS, G. Stephenson and C.W. Kilmister. Concise elegant account for nonspecialists. Lorentz transformation, optical and dynamical applications, more. Bibliography. 108pp. 5⅜ × 8½. 65519-9 Pa. $3.95

INTRODUCTION TO ANALYSIS, Maxwell Rosenlicht. Unusually clear, accessible coverage of set theory, real number system, metric spaces, continuous functions, Riemann integration, multiple integrals, more. Wide range of problems. Undergraduate level. Bibliography. 254pp. 5⅜ × 8½. 65038-3 Pa. $7.00

INTRODUCTION TO QUANTUM MECHANICS With Applications to Chemistry, Linus Pauling & E. Bright Wilson, Jr. Classic undergraduate text by Nobel Prize winner applies quantum mechanics to chemical and physical problems. Numerous tables and figures enhance the text. Chapter bibliographies. Appendices. Index. 468pp. 5⅜ × 8½. 64871-0 Pa. $9.95

ASYMPTOTIC EXPANSIONS OF INTEGRALS, Norman Bleistein & Richard A. Handelsman. Best introduction to important field with applications in a variety of scientific disciplines. New preface. Problems. Diagrams. Tables. Bibliography. Index. 448pp. 5⅜ × 8½. 65082-0 Pa. $10.95

MATHEMATICS APPLIED TO CONTINUUM MECHANICS, Lee A. Segel. Analyzes models of fluid flow and solid deformation. For upper-level math, science and engineering students. 608pp. 5⅜ × 8½. 65369-2 Pa. $12.95

ELEMENTS OF REAL ANALYSIS, David A. Sprecher. Classic text covers fundamental concepts, real number system, point sets, functions of a real variable, Fourier series, much more. Over 500 exercises. 352pp. 5⅜ × 8½. 65385-4 Pa. $8.95

PHYSICAL PRINCIPLES OF THE QUANTUM THEORY, Werner Heisenberg. Nobel Laureate discusses quantum theory, uncertainty, wave mechanics, work of Dirac, Schroedinger, Compton, Wilson, Einstein, etc. 184pp. 5⅜ × 8½. 60113-7 Pa. $4.95

INTRODUCTORY REAL ANALYSIS, A.N. Kolmogorov, S.V. Fomin. Translated by Richard A. Silverman. Self-contained, evenly paced introduction to real and functional analysis. Some 350 problems. 403pp. 5⅜ × 8½. 61226-0 Pa. $7.95

PROBLEMS AND SOLUTIONS IN QUANTUM CHEMISTRY AND PHYSICS, Charles S. Johnson, Jr. and Lee G. Pedersen. Unusually varied problems, detailed solutions in coverage of quantum mechanics, wave mechanics, angular momentum, molecular spectroscopy, scattering theory, more. 280 problems plus 139 supplementary exercises. 430pp. 6½ × 9¼. 65236-X Pa. $10.95

THE ELECTROMAGNETIC FIELD, Albert Shadowitz. Comprehensive undergraduate text covers basics of electric and magnetic fields, builds up to electromagnetic theory. Also related topics, including relativity. Over 900 problems. 768pp. 5⅜ × 8¼. 65660-8 Pa. $15.95

FOURIER SERIES, Georgi P. Tolstov. Translated by Richard A. Silverman. A valuable addition to the literature on the subject, moving clearly from subject to subject and theorem to theorem. 107 problems, answers. 336pp. 5⅜ × 8½. 63317-9 Pa. $7.95

THEORY OF ELECTROMAGNETIC WAVE PROPAGATION, Charles Herach Papas. Graduate-level study discusses the Maxwell field equations, radiation from wire antennas, the Doppler effect and more. xiii + 244pp. 5⅜ × 8½. 65678-0 Pa. $6.95

DISTRIBUTION THEORY AND TRANSFORM ANALYSIS: An Introduction to Generalized Functions, with Applications, A.H. Zemanian. Provides basics of distribution theory, describes generalized Fourier and Laplace transformations. Numerous problems. 384pp. 5⅜ × 8½. 65479-6 Pa. $8.95

THE PHYSICS OF WAVES, William C. Elmore and Mark A. Heald. Unique overview of classical wave theory. Acoustics, optics, electromagnetic radiation, more. Ideal as classroom text or for self-study. Problems. 477pp. 5⅜ × 8½. 64926-1 Pa. $10.95

CALCULUS OF VARIATIONS WITH APPLICATIONS, George M. Ewing. Applications-oriented introduction to variational theory develops insight and promotes understanding of specialized books, research papers. Suitable for advanced undergraduate/graduate students as primary, supplementary text. 352pp. 5⅜ × 8½. 64856-7 Pa. $8.50

A TREATISE ON ELECTRICITY AND MAGNETISM, James Clerk Maxwell. Important foundation work of modern physics. Brings to final form Maxwell's theory of electromagnetism and rigorously derives his general equations of field theory. 1,084pp. 5⅜ × 8½. 60636-8, 60637-6 Pa., Two-vol. set $19.00

AN INTRODUCTION TO THE CALCULUS OF VARIATIONS, Charles Fox. Graduate-level text covers variations of an integral, isoperimetrical problems, least action, special relativity, approximations, more. References. 279pp. 5⅜ × 8½. 65499-0 Pa. $6.95

HYDRODYNAMIC AND HYDROMAGNETIC STABILITY, S. Chandrasekhar. Lucid examination of the Rayleigh-Benard problem; clear coverage of the theory of instabilities causing convection. 704pp. 5⅜ × 8¼. 64071-X Pa. $12.95

CALCULUS OF VARIATIONS, Robert Weinstock. Basic introduction covering isoperimetric problems, theory of elasticity, quantum mechanics, electrostatics, etc. Exercises throughout. 326pp. 5⅜ × 8½. 63069-2 Pa. $7.95

DYNAMICS OF FLUIDS IN POROUS MEDIA, Jacob Bear. For advanced students of ground water hydrology, soil mechanics and physics, drainage and irrigation engineering and more. 335 illustrations. Exercises, with answers. 784pp. 6⅛ × 9¼. 65675-6 Pa. $19.95

NUMERICAL METHODS FOR SCIENTISTS AND ENGINEERS, Richard Hamming. Classic text stresses frequency approach in coverage of algorithms, polynomial approximation, Fourier approximation, exponential approximation, other topics. Revised and enlarged 2nd edition. 721pp. 5⅜ × 8½.
65241-6 Pa. $14.95

THEORETICAL SOLID STATE PHYSICS, Vol. I: Perfect Lattices in Equilibrium; Vol. II: Non-Equilibrium and Disorder, William Jones and Norman H. March. Monumental reference work covers fundamental theory of equilibrium properties of perfect crystalline solids, non-equilibrium properties, defects and disordered systems. Appendices. Problems. Preface. Diagrams. Index. Bibliography. Total of 1,301pp. 5⅜ × 8½. Two volumes. Vol. I 65015-4 Pa. $12.95
Vol. II 65016-2 Pa. $12.95

OPTIMIZATION THEORY WITH APPLICATIONS, Donald A. Pierre. Broad-spectrum approach to important topic. Classical theory of minima and maxima, calculus of variations, simplex technique and linear programming, more. Many problems, examples. 640pp. 5⅜ × 8½. 65205-X Pa. $12.95

THE MODERN THEORY OF SOLIDS, Frederick Seitz. First inexpensive edition of classic work on theory of ionic crystals, free-electron theory of metals and semiconductors, molecular binding, much more. 736pp. 5⅜ × 8½.
65482-6 Pa. $14.95

ESSAYS ON THE THEORY OF NUMBERS, Richard Dedekind. Two classic essays by great German mathematician: on the theory of irrational numbers; and on transfinite numbers and properties of natural numbers. 115pp. 5⅜ × 8½.
21010-3 Pa. $4.95

THE FUNCTIONS OF MATHEMATICAL PHYSICS, Harry Hochstadt. Comprehensive treatment of orthogonal polynomials, hypergeometric functions, Hill's equation, much more. Bibliography. Index. 322pp. 5⅜ × 8½. 65214-9 Pa. $8.95

NUMBER THEORY AND ITS HISTORY, Oystein Ore. Unusually clear, accessible introduction covers counting, properties of numbers, prime numbers, much more. Bibliography. 380pp. 5⅜ × 8½. 65620-9 Pa. $8.95

THE VARIATIONAL PRINCIPLES OF MECHANICS, Cornelius Lanczos. Graduate level coverage of calculus of variations, equations of motion, relativistic mechanics, more. First inexpensive paperbound edition of classic treatise. Index. Bibliography. 418pp. 5⅜ × 8½. 65067-7 Pa. $10.95

MATHEMATICAL TABLES AND FORMULAS, Robert D. Carmichael and Edwin R. Smith. Logarithms, sines, tangents, trig functions, powers, roots, reciprocals, exponential and hyperbolic functions, formulas and theorems. 269pp. 5⅜ × 8½. 60111-0 Pa. $5.95

THEORETICAL PHYSICS, Georg Joos, with Ira M. Freeman. Classic overview covers essential math, mechanics, electromagnetic theory, thermodynamics, quantum mechanics, nuclear physics, other topics. First paperback edition. xxiii + 885pp. 5⅜ × 8½. 65227-0 Pa. $17.95

CATALOG OF DOVER BOOKS

HANDBOOK OF MATHEMATICAL FUNCTIONS WITH FORMULAS, GRAPHS, AND MATHEMATICAL TABLES, edited by Milton Abramowitz and Irene A. Stegun. Vast compendium: 29 sets of tables, some to as high as 20 places. 1,046pp. 8 × 10½. 61272-4 Pa. $21.95

MATHEMATICAL METHODS IN PHYSICS AND ENGINEERING, John W. Dettman. Algebraically based approach to vectors, mapping, diffraction, other topics in applied math. Also generalized functions, analytic function theory, more. Exercises. 448pp. 5⅜ × 8¼. 65649-7 Pa. $8.95

A SURVEY OF NUMERICAL MATHEMATICS, David M. Young and Robert Todd Gregory. Broad self-contained coverage of computer-oriented numerical algorithms for solving various types of mathematical problems in linear algebra, ordinary and partial, differential equations, much more. Exercises. Total of 1,248pp. 5⅜ × 8½. Two volumes. Vol. I 65691-8 Pa. $13.95
Vol. II 65692-6 Pa. $13.95

TENSOR ANALYSIS FOR PHYSICISTS, J.A. Schouten. Concise exposition of the mathematical basis of tensor analysis, integrated with well-chosen physical examples of the theory. Exercises. Index. Bibliography. 289pp. 5⅜ × 8½. 65582-2 Pa. $7.95

INTRODUCTION TO NUMERICAL ANALYSIS (2nd Edition), F.B. Hildebrand. Classic, fundamental treatment covers computation, approximation, interpolation, numerical differentiation and integration, other topics. 150 new problems. 669pp. 5⅜ × 8½. 65363-3 Pa. $13.95

INVESTIGATIONS ON THE THEORY OF THE BROWNIAN MOVEMENT, Albert Einstein. Five papers (1905–8) investigating dynamics of Brownian motion and evolving elementary theory. Notes by R. Fürth. 122pp. 5⅜ × 8½. 60304-0 Pa. $3.95

NUMERICAL METHODS FOR SCIENTISTS AND ENGINEERS, Richard Hamming. Classic text stresses frequency approach in coverage of algorithms, polynomial approximation, Fourier approximation, exponential approximation, other topics. Revised and enlarged 2nd edition. 721pp. 5⅜ × 8½. 65241-6 Pa. $14.95

AN INTRODUCTION TO STATISTICAL THERMODYNAMICS, Terrell L. Hill. Excellent basic text offers wide-ranging coverage of quantum statistical mechanics, systems of interacting molecules, quantum statistics, more. 523pp. 5⅜ × 8½. 65242-4 Pa. $10.95

ELEMENTARY DIFFERENTIAL EQUATIONS, William Ted Martin and Eric Reissner. Exceptionally, clear comprehensive introduction at undergraduate level. Nature and origin of differential equations, differential equations of first, second and higher orders. Picard's Theorem, much more. Problems with solutions. 331pp. 5⅜ × 8½. 65024-3 Pa. $8.95

STATISTICAL PHYSICS, Gregory H. Wannier. Classic text combines thermodynamics, statistical mechanics and kinetic theory in one unified presentation of thermal physics. Problems with solutions. Bibliography. 532pp. 5⅜ × 8½. 65401-X Pa. $10.95

ORDINARY DIFFERENTIAL EQUATIONS, Morris Tenenbaum and Harry Pollard. Exhaustive survey of ordinary differential equations for undergraduates in mathematics, engineering, science. Thorough analysis of theorems. Diagrams. Bibliography. Index. 818pp. 5⅜ × 8½. 64940-7 Pa. $15.95

STATISTICAL MECHANICS: Principles and Applications, Terrell L. Hill. Standard text covers fundamentals of statistical mechanics, applications to fluctuation theory, imperfect gases, distribution functions, more. 448pp. 5⅜ × 8½. 65390-0 Pa. $9.95

ORDINARY DIFFERENTIAL EQUATIONS AND STABILITY THEORY: An Introduction, David A. Sánchez. Brief, modern treatment. Linear equation, stability theory for autonomous and nonautonomous systems, etc. 164pp. 5⅜ × 8¼. 63828-6 Pa. $4.95

THIRTY YEARS THAT SHOOK PHYSICS: The Story of Quantum Theory, George Gamow. Lucid, accessible introduction to influential theory of energy and matter. Careful explanations of Dirac's anti-particles, Bohr's model of the atom, much more. 12 plates. Numerous drawings. 240pp. 5⅜ × 8½. 24895-X Pa. $5.95

ORDINARY DIFFERENTIAL EQUATIONS, I.G. Petrovski. Covers basic concepts, some differential equations and such aspects of the general theory as Euler lines, Arzel's theorem, Peano's existence theorem, Osgood's uniqueness theorem, more. 45 figures. Problems. Bibliography. Index. xi + 232pp. 5⅜ × 8½. 64683-1 Pa. $6.00

GREAT EXPERIMENTS IN PHYSICS: Firsthand Accounts from Galileo to Einstein, edited by Morris H. Shamos. 25 crucial discoveries: Newton's laws of motion, Chadwick's study of the neutron, Hertz on electromagnetic waves, more. Original accounts clearly annotated. 370pp. 5⅜ × 8½. 25346-5 Pa. $8.95

INTRODUCTION TO PARTIAL DIFFERENTIAL EQUATIONS WITH APPLICATIONS, E.C. Zachmanoglou and Dale W. Thoe. Essentials of partial differential equations applied to common problems in engineering and the physical sciences. Problems and answers. 416pp. 5⅜ × 8½. 65251-3 Pa. $9.95

BURNHAM'S CELESTIAL HANDBOOK, Robert Burnham, Jr. Thorough guide to the stars beyond our solar system. Exhaustive treatment. Alphabetical by constellation: Andromeda to Cetus in Vol. 1; Chamaeleon to Orion in Vol. 2; and Pavo to Vulpecula in Vol. 3. Hundreds of illustrations. Index in Vol. 3. 2,000pp. 6⅛ × 9¼. 23567-X, 23568-8, 23673-0 Pa., Three-vol. set $38.85

ASYMPTOTIC EXPANSIONS FOR ORDINARY DIFFERENTIAL EQUATIONS, Wolfgang Wasow. Outstanding text covers asymptotic power series, Jordan's canonical form, turning point problems, singular perturbations, much more. Problems. 384pp. 5⅜ × 8½. 65456-7 Pa. $8.95

AMATEUR ASTRONOMER'S HANDBOOK, J.B. Sidgwick. Timeless, comprehensive coverage of telescopes, mirrors, lenses, mountings, telescope drives, micrometers, spectroscopes, more. 189 illustrations. 576pp. 5⅜ × 8¼. 24034-7 Pa. $8.95

ROTARY-WING AERODYNAMICS, W.Z. Stepniewski. Clear, concise text covers aerodynamic phenomena of the rotor and offers guidelines for helicopter performance evaluation. Originally prepared for NASA. 537 figures. 640pp. 6⅛ × 9¼.
64647-5 Pa. $14.95

DIFFERENTIAL GEOMETRY, Heinrich W. Guggenheimer. Local differential geometry as an application of advanced calculus and linear algebra. Curvature, transformation groups, surfaces, more. Exercises. 62 figures. 378pp. 5⅜ × 8½.
63433-7 Pa. $7.95

INTRODUCTION TO SPACE DYNAMICS, William Tyrrell Thomson. Comprehensive, classic introduction to space-flight engineering for advanced undergraduate and graduate students. Includes vector algebra, kinematics, transformation of coordinates. Bibliography. Index. 352pp. 5⅜ × 8½.   65113-4 Pa. $8.00

A SURVEY OF MINIMAL SURFACES, Robert Osserman. Up-to-date, in-depth discussion of the field for advanced students. Corrected and enlarged edition covers new developments. Includes numerous problems. 192pp. 5⅜ × 8½.
64998-9 Pa. $8.00

ANALYTICAL MECHANICS OF GEARS, Earle Buckingham. Indispensable reference for modern gear manufacture covers conjugate gear-tooth action, gear-tooth profiles of various gears, many other topics. 263 figures. 102 tables. 546pp. 5⅜ × 8½.   65712-4 Pa. $11.95

SET THEORY AND LOGIC, Robert R. Stoll. Lucid introduction to unified theory of mathematical concepts. Set theory and logic seen as tools for conceptual understanding of real number system. 496pp. 5⅜ × 8¼.   63829-4 Pa. $8.95

A HISTORY OF MECHANICS, René Dugas. Monumental study of mechanical principles from antiquity to quantum mechanics. Contributions of ancient Greeks, Galileo, Leonardo, Kepler, Lagrange, many others. 671pp. 5⅜ × 8½.
65632-2 Pa. $14.95

FAMOUS PROBLEMS OF GEOMETRY AND HOW TO SOLVE THEM, Benjamin Bold. Squaring the circle, trisecting the angle, duplicating the cube: learn their history, why they are impossible to solve, then solve them yourself. 128pp. 5⅜ × 8½.   24297-8 Pa. $3.95

MECHANICAL VIBRATIONS, J.P. Den Hartog. Classic textbook offers lucid explanations and illustrative models, applying theories of vibrations to a variety of practical industrial engineering problems. Numerous figures. 233 problems, solutions. Appendix. Index. Preface. 436pp. 5⅜ × 8½.   64785-4 Pa. $8.95

CURVATURE AND HOMOLOGY, Samuel I. Goldberg. Thorough treatment of specialized branch of differential geometry. Covers Riemannian manifolds, topology of differentiable manifolds, compact Lie groups, other topics. Exercises. 315pp. 5⅜ × 8½.   64314-X Pa. $6.95

HISTORY OF STRENGTH OF MATERIALS, Stephen P. Timoshenko. Excellent historical survey of the strength of materials with many references to the theories of elasticity and structure. 245 figures. 452pp. 5⅜ × 8½.   61187-6 Pa. $9.95

GEOMETRY OF COMPLEX NUMBERS, Hans Schwerdtfeger. Illuminating, widely praised book on analytic geometry of circles, the Moebius transformation, and two-dimensional non-Euclidean geometries. 200pp. 5⅜ × 8¼.
63830-8 Pa. $6.95

MECHANICS, J.P. Den Hartog. A classic introductory text or refresher. Hundreds of applications and design problems illuminate fundamentals of trusses, loaded beams and cables, etc. 334 answered problems. 462pp. 5⅜ × 8½. 60754-2 Pa. $8.95

TOPOLOGY, John G. Hocking and Gail S. Young. Superb one-year course in classical topology. Topological spaces and functions, point-set topology, much more. Examples and problems. Bibliography. Index. 384pp. 5⅜ × 8¼.
65676-4 Pa. $7.95

STRENGTH OF MATERIALS, J.P. Den Hartog. Full, clear treatment of basic material (tension, torsion, bending, etc.) plus advanced material on engineering methods, applications. 350 answered problems. 323pp. 5⅜ × 8½. 60755-0 Pa. $7.50

ELEMENTARY CONCEPTS OF TOPOLOGY, Paul Alexandroff. Elegant, intuitive approach to topology from set-theoretic topology to Betti groups; how concepts of topology are useful in math and physics. 25 figures. 57pp. 5⅜ × 8½.
60747-X Pa. $2.95

ADVANCED STRENGTH OF MATERIALS, J.P. Den Hartog. Superbly written advanced text covers torsion, rotating disks, membrane stresses in shells, much more. Many problems and answers. 388pp. 5⅜ × 8½. 65407-9 Pa. $8.95

COMPUTABILITY AND UNSOLVABILITY, Martin Davis. Classic graduate-level introduction to theory of computability, usually referred to as theory of recurrent functions. New preface and appendix. 288pp. 5⅜ × 8½. 61471-9 Pa. $6.95

GENERAL CHEMISTRY, Linus Pauling. Revised 3rd edition of classic first-year text by Nobel laureate. Atomic and molecular structure, quantum mechanics, statistical mechanics, thermodynamics correlated with descriptive chemistry. Problems. 992pp. 5⅜ × 8½. 65622-5 Pa. $18.95

AN INTRODUCTION TO MATRICES, SETS AND GROUPS FOR SCIENCE STUDENTS, G. Stephenson. Concise, readable text introduces sets, groups, and most importantly, matrices to undergraduate students of physics, chemistry, and engineering. Problems. 164pp. 5⅜ × 8½. 65077-4 Pa. $5.95

THE HISTORICAL BACKGROUND OF CHEMISTRY, Henry M. Leicester. Evolution of ideas, not individual biography. Concentrates on formulation of a coherent set of chemical laws. 260pp. 5⅜ × 8½. 61053-5 Pa. $6.00

THE PHILOSOPHY OF MATHEMATICS: An Introductory Essay, Stephan Körner. Surveys the views of Plato, Aristotle, Leibniz & Kant concerning propositions and theories of applied and pure mathematics. Introduction. Two appendices. Index. 198pp. 5⅜ × 8½. 25048-2 Pa. $5.95

THE DEVELOPMENT OF MODERN CHEMISTRY, Aaron J. Ihde. Authoritative history of chemistry from ancient Greek theory to 20th-century innovation. Covers major chemists and their discoveries. 209 illustrations. 14 tables. Bibliographies. Indices. Appendices. 851pp. 5⅜ × 8½. 64235-6 Pa. $15.95

THE FOUR-COLOR PROBLEM: Assaults and Conquest, Thomas L. Saaty and Paul G. Kainen. Engrossing, comprehensive account of the century-old combinatorial topological problem, its history and solution. Bibliographies. Index. 110 figures. 228pp. 5⅜ × 8½. 65092-8 Pa. $6.00

CATALYSIS IN CHEMISTRY AND ENZYMOLOGY, William P. Jencks. Exceptionally clear coverage of mechanisms for catalysis, forces in aqueous solution, carbonyl- and acyl-group reactions, practical kinetics, more. 864pp. 5⅜ × 8½. 65460-5 Pa. $18.95

PROBABILITY: An Introduction, Samuel Goldberg. Excellent basic text covers set theory, probability theory for finite sample spaces, binomial theorem, much more. 360 problems. Bibliographies. 322pp. 5⅜ × 8½. 65252-1 Pa. $7.95

LIGHTNING, Martin A. Uman. Revised, updated edition of classic work on the physics of lightning. Phenomena, terminology, measurement, photography, spectroscopy, thunder, more. Reviews recent research. Bibliography. Indices. 320pp. 5⅜ × 8¼. 64575-4 Pa. $7.95

PROBABILITY THEORY: A Concise Course, Y.A. Rozanov. Highly readable, self-contained introduction covers combination of events, dependent events, Bernoulli trials, etc. Translation by Richard Silverman. 148pp. 5⅜ × 8¼. 63544-9 Pa. $4.50

THE CEASELESS WIND: An Introduction to the Theory of Atmospheric Motion, John A. Dutton. Acclaimed text integrates disciplines of mathematics and physics for full understanding of dynamics of atmospheric motion. Over 400 problems. Index. 97 illustrations. 640pp. 6 × 9. 65096-0 Pa. $16.95

STATISTICS MANUAL, Edwin L. Crow, et al. Comprehensive, practical collection of classical and modern methods prepared by U.S. Naval Ordnance Test Station. Stress on use. Basics of statistics assumed. 288pp. 5⅜ × 8½. 60599-X Pa. $6.00

WIND WAVES: Their Generation and Propagation on the Ocean Surface, Blair Kinsman. Classic of oceanography offers detailed discussion of stochastic processes and power spectral analysis that revolutionized ocean wave theory. Rigorous, lucid. 676pp. 5⅜ × 8½. 64652-1 Pa. $14.95

STATISTICAL METHOD FROM THE VIEWPOINT OF QUALITY CONTROL, Walter A. Shewhart. Important text explains regulation of variables, uses of statistical control to achieve quality control in industry, agriculture, other areas. 192pp. 5⅜ × 8½. 65232-7 Pa. $6.00

THE INTERPRETATION OF GEOLOGICAL PHASE DIAGRAMS, Ernest G. Ehlers. Clear, concise text emphasizes diagrams of systems under fluid or containing pressure; also coverage of complex binary systems, hydrothermal melting, more. 288pp. 6½ × 9¼. 65389-7 Pa. $8.95

STATISTICAL ADJUSTMENT OF DATA, W. Edwards Deming. Introduction to basic concepts of statistics, curve fitting, least squares solution, conditions without parameter, conditions containing parameters. 26 exercises worked out. 271pp. 5⅜ × 8½. 64685-8 Pa. $7.95

CATALOG OF DOVER BOOKS

DE RE METALLICA, Georgius Agricola. The famous Hoover translation of greatest treatise on technological chemistry, engineering, geology, mining of early modern times (1556). All 289 original woodcuts. 638pp. 6¾ × 11.
60006-8 Clothbd. $15.95

SOME THEORY OF SAMPLING, William Edwards Deming. Analysis of the problems, theory and design of sampling techniques for social scientists, industrial managers and others who find statistics increasingly important in their work. 61 tables. 90 figures. xvii + 602pp. 5⅜ × 8½.
64684-X Pa. $14.95

THE VARIOUS AND INGENIOUS MACHINES OF AGOSTINO RAMELLI: A Classic Sixteenth-Century Illustrated Treatise on Technology, Agostino Ramelli. One of the most widely known and copied works on machinery in the 16th century. 194 detailed plates of water pumps, grain mills, cranes, more. 608pp. 9 × 12.
25497-6 Clothbd. $34.95

LINEAR PROGRAMMING AND ECONOMIC ANALYSIS, Robert Dorfman, Paul A. Samuelson and Robert M. Solow. First comprehensive treatment of linear programming in standard economic analysis. Game theory, modern welfare economics, Leontief input-output, more. 525pp. 5⅜ × 8½.
65491-5 Pa. $12.95

ELEMENTARY DECISION THEORY, Herman Chernoff and Lincoln E. Moses. Clear introduction to statistics and statistical theory covers data processing, probability and random variables, testing hypotheses, much more. Exercises. 364pp. 5⅜ × 8½.
65218-1 Pa. $8.95

THE COMPLEAT STRATEGYST: Being a Primer on the Theory of Games of Strategy, J.D. Williams. Highly entertaining classic describes, with many illustrated examples, how to select best strategies in conflict situations. Prefaces. Appendices. 268pp. 5⅜ × 8½.
25101-2 Pa. $5.95

MATHEMATICAL METHODS OF OPERATIONS RESEARCH, Thomas L. Saaty. Classic graduate-level text covers historical background, classical methods of forming models, optimization, game theory, probability, queueing theory, much more. Exercises. Bibliography. 448pp. 5⅜ × 8¼.
65703-5 Pa. $12.95

CONSTRUCTIONS AND COMBINATORIAL PROBLEMS IN DESIGN OF EXPERIMENTS, Damaraju Raghavarao. In-depth reference work examines orthogonal Latin squares, incomplete block designs, tactical configuration, partial geometry, much more. Abundant explanations, examples. 416pp. 5⅜ × 8¼.
65685-3 Pa. $10.95

THE ABSOLUTE DIFFERENTIAL CALCULUS (CALCULUS OF TENSORS), Tullio Levi-Civita. Great 20th-century mathematician's classic work on material necessary for mathematical grasp of theory of relativity. 452pp. 5⅜ × 8½.
63401-9 Pa. $9.95

VECTOR AND TENSOR ANALYSIS WITH APPLICATIONS, A.I. Borisenko and I.E. Tarapov. Concise introduction. Worked-out problems, solutions, exercises. 257pp. 5⅜ × 8¼.
63833-2 Pa. $6.95

CATALOG OF DOVER BOOKS

TENSOR CALCULUS, J.L. Synge and A. Schild. Widely used introductory text covers spaces and tensors, basic operations in Riemannian space, non-Riemannian spaces, etc. 324pp. 5⅜ × 8¼. 63612-7 Pa. $7.00

A CONCISE HISTORY OF MATHEMATICS, Dirk J. Struik. The best brief history of mathematics. Stresses origins and covers every major figure from ancient Near East to 19th century. 41 illustrations. 195pp. 5⅜ × 8½. 60255-9 Pa. $7.95

A SHORT ACCOUNT OF THE HISTORY OF MATHEMATICS, W.W. Rouse Ball. One of clearest, most authoritative surveys from the Egyptians and Phoenicians through 19th-century figures such as Grassman, Galois, Riemann. Fourth edition. 522pp. 5⅜ × 8½. 20630-0 Pa. $9.95

HISTORY OF MATHEMATICS, David E. Smith. Non-technical survey from ancient Greece and Orient to late 19th century; evolution of arithmetic, geometry, trigonometry, calculating devices, algebra, the calculus. 362 illustrations. 1,355pp. 5⅜ × 8½. 20429-4, 20430-8 Pa., Two-vol. set $21.90

THE GEOMETRY OF RENÉ DESCARTES, René Descartes. The great work founded analytical geometry. Original French text, Descartes' own diagrams, together with definitive Smith-Latham translation. 244pp. 5⅜ × 8½. 60068-8 Pa. $6.00

THE ORIGINS OF THE INFINITESIMAL CALCULUS, Margaret E. Baron. Only fully detailed and documented account of crucial discipline: origins; development by Galileo, Kepler, Cavalieri; contributions of Newton, Leibniz, more. 304pp. 5⅜ × 8½. (Available in U.S. and Canada only) 65371-4 Pa. $7.95

THE HISTORY OF THE CALCULUS AND ITS CONCEPTUAL DEVELOPMENT, Carl B. Boyer. Origins in antiquity, medieval contributions, work of Newton, Leibniz, rigorous formulation. Treatment is verbal. 346pp. 5⅜ × 8½. 60509-4 Pa. $6.95

THE THIRTEEN BOOKS OF EUCLID'S ELEMENTS, translated with introduction and commentary by Sir Thomas L. Heath. Definitive edition. Textual and linguistic notes, mathematical analysis. 2500 years of critical commentary. Not abridged. 1,414pp. 5⅜ × 8½. 60088-2, 60089-0, 60090-4 Pa., Three-vol. set $26.85

A HISTORY OF VECTOR ANALYSIS: The Evolution of the Idea of a Vectorial System, Michael J. Crowe. The first large-scale study of the history of vector analysis, now the standard on the subject. Unabridged republication of the edition published by University of Notre Dame Press, 1967, with second preface by Michael C. Crowe. Index. 278pp. 5⅜ × 8½. 64955-5 Pa. $7.00

THE HISTORICAL ROOTS OF ELEMENTARY MATHEMATICS, Lucas N.H. Bunt, Phillip S. Jones, and Jack D. Bedient. Fundamental underpinnings of modern arithmetic, algebra, geometry and number systems derived from ancient civilizations. 320pp. 5⅜ × 8½. 25563-8 Pa. $7.95

CALCULUS REFRESHER FOR TECHNICAL PEOPLE, A. Albert Klaf. Covers important aspects of integral and differential calculus via 756 questions. 566 problems, most answered. 431pp. 5⅜ × 8½. 20370-0 Pa. $7.95

CATALOG OF DOVER BOOKS

CHALLENGING MATHEMATICAL PROBLEMS WITH ELEMENTARY SOLUTIONS, A.M. Yaglom and I.M. Yaglom. Over 170 challenging problems on probability theory, combinatorial analysis, points and lines, topology, convex polygons, many other topics. Solutions. Total of 445pp. 5⅜ × 8½. Two-vol. set.
Vol. I 65536-9 Pa. $5.95
Vol. II 65537-7 Pa. $5.95

FIFTY CHALLENGING PROBLEMS IN PROBABILITY WITH SOLU-TIONS, Frederick Mosteller. Remarkable puzzlers, graded in difficulty, illustrate elementary and advanced aspects of probability. Detailed solutions. 88pp. 5⅜ × 8½.
65355-2 Pa. $3.95

EXPERIMENTS IN TOPOLOGY, Stephen Barr. Classic, lively explanation of one of the byways of mathematics. Klein bottles, Moebius strips, projective planes, map coloring, problem of the Koenigsberg bridges, much more, described with clarity and wit. 43 figures. 210pp. 5⅜ × 8½.
25933-1 Pa. $4.95

RELATIVITY IN ILLUSTRATIONS, Jacob T. Schwartz. Clear non-technical treatment makes relativity more accessible than ever before. Over 60 drawings illustrate concepts more clearly than text alone. Only high school geometry needed. Bibliography. 128pp. 6⅛ × 9¼.
25965-X Pa. $5.95

AN INTRODUCTION TO ORDINARY DIFFERENTIAL EQUATIONS, Earl A. Coddington. A thorough and systematic first course in elementary differential equations for undergraduates in mathematics and science, with many exercises and problems (with answers). Index. 304pp. 5⅜ × 8¼.
65942-9 Pa. $7.95

FOURIER SERIES AND ORTHOGONAL FUNCTIONS, Harry F. Davis. An incisive text combining theory and practical example to introduce Fourier series, orthogonal functions and applications of the Fourier method to boundary-value problems. 570 exercises. Answers and notes. 416pp. 5⅜ × 8½.
65973-9 Pa. $8.95

THE THEORY OF BRANCHING PROCESSES, Theodore E. Harris. First systematic, comprehensive treatment of branching (i.e. multiplicative) processes and their applications. Galton-Watson model, Markov branching processes, electron-photon cascade, many other topics. Rigorous proofs. Bibliography. 240pp. 5⅜ × 8½.
65952-6 Pa. $6.95

AN INTRODUCTION TO ALGEBRAIC STRUCTURES, Joseph Landin. Superb self-contained text covers "abstract algebra": sets and numbers, theory of groups, theory of rings, much more. Numerous well-chosen examples, exercises. 247pp. 5⅜ × 8½.
65940-2 Pa. $6.95

GAMES AND DECISIONS: Introduction and Critical Survey, R. Duncan Luce and Howard Raiffa. Superb non-technical introduction to game theory, primarily applied to social sciences. Utility theory, zero-sum games, n-person games, decision-making, much more. Bibliography. 509pp. 5⅜ × 8½. 65943-7 Pa. $10.95

Prices subject to change without notice.
Available at your book dealer or write for free Mathematics and Science Catalog to Dept. GI, Dover Publications, Inc., 31 East 2nd St., Mineola, N.Y. 11501. Dover publishes more than 175 books each year on science, elementary and advanced mathematics, biology, music, art, literary history, social sciences and other areas.